Confronting Mistakes

Confronting Mistakes

Lessons from the Aviation Industry when Dealing with Error

Jan U. Hagen

First published 2013 by
PALGRAVE MACMILLAN

Palgrave Macmillan in the UK is an imprint of Macmillan Publishers Limited, registered in England, company number 785998, of Houndmills, Basingstoke, Hampshire RG21 6XS.

Palgrave Macmillan in the US is a division of St Martin's Press LLC, 175 Fifth Avenue, New York, NY 10010.

Palgrave Macmillan is the global academic imprint of the above companies and has companies and representatives throughout the world.

Palgrave® and Macmillan® are registered trademarks in the United States, the United Kingdom, Europe and other countries

ISBN: 978–1–137–27617–9   hardback

This book is printed on paper suitable for recycling and made from fully managed and sustained forest sources. Logging, pulping and manufacturing processes are expected to conform to the environmental regulations of the country of origin.

A catalogue record for this book is available from the British Library.

A catalog record for this book is available from the Library of Congress.

Contents

Foreword

By Captain Ralf Nagel

Confronting Mistakes is a book for managers describing the error management in the cockpits of commercial aircraft. I am not talking about a code of conduct along the lines of "what managers can learn from pilots." Rather, *Confronting Mistakes* is an exciting book that deals with the consequences resulting from a lack of error management. Some of the cases are hair-raising; others shocking yet foreseeable. All had fatal consequences – for those who caused them as well as for bystanders. Yet with the help of a functioning error management, these consequences could have been avoided. This realization, recognized 30 years ago in aviation, eventually led to Crew Resource Management, or CRM. Introduced in the early 1980s, it has by now become well established in the aviation industry, which has continually worked on improving it to make those in the cockpit *proactively* aware of potential sources of error.

Now is the time to introduce the knowledge we have gained since then to a wider audience. After all, the aviation industry, for obvious reasons, has learned error management earlier and more systematically than other industries. Accusations and sanctions have no place therein – the only concern is with a root-cause analysis to prevent the next accident or disaster.

Today, CRM is both an integral part of pilot education as well as regular training. It is practiced worldwide in the cockpit, which also means that the implementation of modern error management is possible, even on the largest scale.

Confronting Mistakes offers fascinating reading for a broad audience and I hope that its pages provide both motivation and inspiration.

Acknowledgments

The realization of this book would have been impossible without the support of multiple individuals. Above all, they gave me a deeper insight into the workings of the cockpits of commercial aircrafts. In this context, I owe a great deal of gratitude to captain Ralf Nagel. Since the beginning of this project, he has supported my efforts and made himself available to me as a critical sounding board. I am indebted to him for letting me fly as an observer in the cockpits of passenger planes and for being able to attend flight simulator training sessions. Thanks to Line Oriented Flight Training, I could see the cooperation of aircrews under varying conditions.

I would also like to thank the many pilots who accepted me in their cockpits, for both their patience and their frankness and want to extend my special thanks to captains Sven Behmer, Stefan Bürgers, Stefan Gilbert, Robin Müller, Asger Rogalla, Ulrich Spielmann, Andreas Spors, Marc Peter Störl, and Thomas Wilpert. To captain Bill Records, I owe a big thank you for helping me describe the crash of United Airlines 232 and to captain Robert Bragg for his help in understanding the Tenerife accident. Many thanks also to Staff Captain Stephan Wrede, who brought me closer to the complex relationships in the cockpits of military aircraft. For their information on the relevance of error management in business, I am particularly grateful to Carl L. von Boehm-Bezing, Manfred Gentz, Kay Kratky, and Ulrich Schürenkrämer. I thank my academic colleagues Derek Abell, Matthew Bothner, Erika Hayes James, Tony Kern, Zhike Lei, Wulff Plinke, and Mary Waller for their ideas. A special thank you goes to Amy C. Edmondson, whose work inspired me to write this book. I am also indebted to the late J. Richard Hackman for his encouragement to take up Crew Resource Management as a future part of organizational behavior. I thank my colleague Olaf Plötner for his long and varied support for the realization of this project. I thank Lena Mahr, Felix Schwalbe, and Inka Warscheid for their help in preparing the graphs and Jeff Reich and Robert Furlong for translation and proofreading. My special thanks go to Gabriele Weber-Jarić, who continuously encouraged me not only to realize this book, but who also repeatedly opened up new perspectives for me and thus gave the project a decisive impetus forward at key points. In addition, she kept a constant eye on the text to see to its readability. I am also indebted to Charles Perrow and Scott A. Snook for their books *Normal Accidents* and *Friendly Fire*, both of which enlightened and inspired me.

This book is dedicated to my wife, Tamara, whom I thank for her tremendous understanding for my undertaking and even more for her critical feedback; and to my son, Chrysander, who repeatedly insisted that I am not allowed to write a boring book.

Part I Pre Crew Resource Management

The crisis on the financial markets began to unfold with increasing severity starting in 2007. Many were left dumbfounded that big-name banks had taken such disproportionately high risks with their structured securities. People saw the remuneration system in the investment banking sector, enormous bonus payouts, and the associated asymmetric risk distribution as the main causes of the crisis. They started asking how things could have spiraled so far out of control. Yet even before the crisis, some parties within the banks had urged caution. The question is why these warnings went unheard. Were they overlooked? Underestimated? What mistakes were made, how did they come about, who failed to pick up on them, and how were they allowed to trigger a series of further errors that ultimately culminated in such dramatic consequences?

The bankers' behavior, however, is by no means an exception. Indeed, it is not even confined to the banking industry. Take Enron, for example, where for a long time no one moved to curb the company's actions even though its management policy breached legislation.[1] Or take BP: in the period before the accident at the Deepwater Horizon oil-drilling rig, it has been reported that there were warnings about inadequate cementing around the drill hole. The same applies to the jamming gas pedals fitted in a range of passenger cars made by Toyota. The faulty models entered the market almost as if designers, automotive engineers, or quality control processes for suppliers were not in place. There have been mistakes, errors, poor decisions, infringements, affairs, and scandals in any organization you care to mention. None of them, though, appears to have had any effective controls in place to intervene in time to prevent things going awry. Instead, those involved could only watch as fate ran its course.

---

1 Hamilton, S. (2004); Oppel, R.A. (2002).

Let us take a look at normal day-to-day operations in a company. What happens if someone makes a mistake or takes the wrong decision? The issue here is not intentional misconduct, fraudulent behavior, gross negligence, or large-scale mismanagement. I am simply referring to the little mistakes, errors, and poor decisions that occur every single day. We are not necessarily even aware of these blunders while making them. In fact, according to a 1979 study by the National Aeronautics and Space Administration (NASA), we make a minor error every four minutes.[2] The original study focused on pilots – a group that, given the risks inherent in their chosen profession, makes every effort to avoid errors. But for them, as for us, errors are the result of momentary blackouts, a temporary short circuit in the brain, false impressions, deceptive memories, dots wrongly combined, fragments of conversation that we interpret incorrectly, prejudices – be they sexual, social, or ideological – momentary feelings of mental imbalance, disorientation, stress, and other disturbances. Or, as Charles Perrow described it in his brilliant work titled *Normal Accidents*, we "zig, when we should have zagged, even when we are attentive and can see."[3] At best, we may have a hunch that what we do or believe or say is not 100 percent solid. Perrow suggests that one of the reasons for this kind of behavior is "that we construct an expected world because we can't handle the complexity of the present one, and then process the information that fits the expected world, and find reasons to exclude the information that might contradict it."[4] The fact is that we draw conclusions, derive knowledge, and make judgments based on mere assumptions. We see connections and relationships where there are only tenuous links. We misunderstand contexts and take stories and information as objective facts when they are actually hearsay. Basically, our problem is that we believe we can and should be "right," when in reality we start out with "quasi-right" at best and ideally adjust our decisions and actions as we proceed. The alternative – that is, believing that we are right and later realizing that we were wrong – creates a state of confusion leading to uncomfortable questions as to the validity of our convictions per se. If I say, "I know my car keys are in the upper drawer of the little table in the entrance hall. No question about it. I have been putting them there every day for ages," it will not sit nicely with me if I suddenly realize I have put them into the bedside locker. Actually, it will be so upsetting that I will force myself to repeatedly reconstruct how the keys ended up in the bedside locker, just to reclaim the sense that I am living in an ordered, plausible world.

---

2 Ruffell Smith, H.P. (1979), pp. 14–21.
3 Perrow, C. (1999), p. 214.
4 Ibid., p. 214.

In the daily business environment, these apparently rock-solid percep-
tions are all too often generated under pressure – namely, the pressure of
deadlines, of getting a job done, and the pressure to succeed. This, too,
renders their validity as being no more than relative. In concrete terms,
this can result in incorrect figures that feed into decision-making processes. This in turn can lead to skewed interest rates, market shares, and
growth rates. Alternatively, we might encounter causalities that – although
derived from statistical analyses that are only *partially* complete – nevertheless mutate into the foundations used for key business decisions.[5] Of
course, there are actions that fail even though all the necessary conditions
were apparently in place. It does not mean that the people involved were
stupid or thought it would be fun to be reckless for a change. To quote
Charles Perrow again, "in fact, very few people in any walk of life deserve
that appellation [of being stupid]. Nor does the attribute 'risk taker' help
us much [...] As drivers, we all would probably admit that at times we
took unnecessary risks; but what we say to ourselves and others is, 'I don't
know why; it was silly, stupid of me.' We generally do not do it because it
was exciting."[6]

Yet, all these examples deal with errors that – if we took the time to analyze them properly – we could learn from, or at least use to make us more
aware. We might want to find out why they occurred and work out how to
prevent them from happening again. Managers in particular should surely
be interested in doing just that.

A year ago, my ESMT colleagues and I wanted to know whether this is
the case: do managers discuss errors made by their employees? Do they step
in if a member of staff makes a mistake when calculating a rival company's
market share in a competitor analysis? Yes, we found out, most of them do.
But do employees also say something if their superior gets his or her figures
wrong or looks set to make a questionable decision? Here, as we learned,
people are far more reluctant to speak up.[7]

One problem is that errors are often associated with shoddy work, failure, and personal weakness. That is why we find it unpleasant to discuss
them openly. However, mistakes are not necessarily a result of carelessness
or a lack of skill and ability. A person may be distracted, tired, or overbur-
dened. Personal issues are by no means the only possible reasons for this.
The problem may, in fact, lie in the work environment.

---

5 Reason, J. (1997), pp. 71–72.
6 Perrow, C. (1999), p. 214.
7 This is in line with previous research of Edmondson, A.C. (1996) and Milliken, F.J. et al.
   (2003).

To get to the bottom of things, we asked 360 managers from differ-ent sectors about how they deal with errors. We looked at their willing-ness to discuss mistakes made by others: top-down, bottom-up, and among direct colleagues. We also asked them to evaluate to what extent errors are accepted within their own corporate culture. Some 41 percent of those questioned worked in companies with more than 10,000 employees. The managers' average age was 43, and they were in charge of an average of 150 employees with a minimum of eight. Of those questioned, 83 percent were middle management and 11 percent were owners, CEOs, or manag-ing directors. Women accounted for 12 percent.

How do managers address errors made by employees, colleagues, and superiors? Our survey revealed that if they discovered an error made by an employee or colleague, 88 percent of managers would raise the issue privately behind closed doors, 11 percent would discuss it openly, and just 1 percent would ignore the error. When it comes to pointing out a mistake made by someone higher up the ranks, 86 percent would do so in private. Only 4 percent would be prepared to broach the issue openly, whereas 10 percent would rather keep any knowledge of an error made by their superior to themselves.

How do employees, colleagues, and superiors report errors? We asked managers how their own employees, colleagues, and superiors speak to them about errors. Just 54 percent said they would mainly do so in pri-vate; 18 percent said mistakes were addressed in a more open forum. A further 28 percent assumed they were never actually made aware of their mistakes. However, these figures do not tally with the previous results. Of those questioned, 88 percent claimed that they themselves would generally address errors made by others in private. Yet only 54 percent believed they are being informed of errors in this way. In contrast with the 11 percent quoted earlier, 18 percent said their own errors are discussed in front of people. That could be because this experience has stuck in their minds more than those occasions in which they addressed others' mistakes in a public forum.

In this context Kathryn Schulz reminds us of one of Freud's spectacu-lar errors. "Once, while settling his monthly accounts, Freud came upon a name of a patient whose case history he couldn't recall, even though he could see that he had visited her every day for many weeks, scarcely six months previously. He tried for a long time to bring the patient to mind, but for the life of him was unable to do so. When the memory finally came back to him, Freud was astonished by his 'almost incredible instance of forgetting.' The patient in question was a young woman whose parents brought her in because she complained incessantly of stomach pains. Freud diagnosed her with hysteria." A few months later, she died of cancer.

"On the whole," Schulz concluded, "our ability to forget our mistakes seems keener than our ability to remember them."[8] Intuitively we might see this differently and feel that mistakes are so embarrassing that we will never be able to forget them, particularly if they were large. The answer to that is that both reactions coexist. We are able to either suppress the memory of our mistakes or remember them painfully.

Quite fittingly, 28 percent of the managers in our survey fear that discussions about their mistakes take place behind their backs. Alternatively, they hope – quite unrealistically – that no one will ever notice their mistakes. Given that we may be unaware while erring, we can be sure that others are not, as we usually notice other people's errancies quite early on. The notion that we can successfully hide our errors or mistakes is one of our managers' – and everybody else's – many instances of wishful thinking and hoping against hope. "What with our error-blindness," Schulz writes, "our amnesia for our mistakes, the lack of a category called 'error,' [...] it's no wonder we have so much trouble accepting that wrongness is a part of who we are. Because we don't experience, remember, track, or retain mistakes as a feature of our inner landscape, wrongness always seems to come at us from left field – that is, from outside ourselves. But the reality could hardly be more different. Error is the ultimate inside job."[9]

Male and female managers. When it comes to the error management of male and female managers, both sexes prefer to speak to their superiors about an error in private. However, this tendency was slightly less pronounced among the female managers (78 percent compared to 87 percent of male managers). In contrast, female managers were almost three times more willing than their male colleagues to discuss errors openly; only 10 percent of those questioned admitted that they would never openly address errors made by their superiors. Sixty-seven percent of the women in this group were instead more likely to talk about a mistake made by their superior behind his or her back. Only 39 percent of the male managers would behave in this way, with the majority opting to keep silent.

Younger and older managers. A comparison of the various age groups revealed the biggest differences. The older the manager, the more he or she is willing to speak to employees and superiors about errors. Unsurprisingly, younger managers are more cautious. Fifteen percent of those under 30

---

8 Schulz, K. (2010), p. 19.
9 Ibid., p. 21.

would keep a superior's error to themselves. All of the younger managers were keen for errors to be discussed in private. Only 71 percent of managers over 60 felt the same way. In addition, the younger managers were more reluctant about their own errors being addressed openly, with nearly 20 percent saying they would prefer to correct their errors discretely by themselves. Here, too, there is a clear shift in attitude as managers get older, with only 6 percent of the over-60s sharing this view. What is more, this behavior does not depend on the manager's position within the hierarchy. Older managers in the lower and middle ranks proved more willing to discuss errors than younger managers further up the hierarchy.

How do managers rate their corporate error culture? According to our survey, errors are perceived to be part of a company's normal operations. Of those questioned, 75 percent agreed with this view, with just 25 percent believing that errors were the result of careless work, associating them with a feeling of embarrassment and/or sanctions. In the workplace of those questioned, the modern forms of error management applied in the airline industry were nowhere to be found.

What does this mean for competitive companies? Today's global corporations increasingly rely on networked processes supported by cutting-edge communication technologies. To resolve problems promptly, these types of systems demand open discussions that also focus on the cause of errors. This not only refers to technical breakdowns and damage caused by third parties, but also covers mistakes, bad decisions, blunders, and errors made by those directly involved. These discussions must be rational and analytical because attributing blame, becoming defensive, and stirring up emotions will only slow down or hamper operations.

No doubt, most companies still have a long way to go before error management becomes a regular part of day-to-day work life. The first step has been made, though – according to our study, most managers accept errors as being a normal part of the work culture. There is only one aspect that does not fit, namely, the overwhelming preference for discussing errors in private and involving as few people as possible. It is a behavior still associated with shame and embarrassment. That does not even begin to address the alternative behavior of remaining silent or whispering in secret.

Yet, active error management can work and be successful. A look at high-risk industries such as aviation shows that professionals in these fields have come to openly accept and analyze errors before eliminating the causes. This matter-of-fact approach is essential in avoiding disasters, because any errors in the passenger transportation sector can have catastrophic results, so clearly and openly that it is simply impossible to sugar-coat them or brush them under the carpet. For that reason alone – and unlike bad investments, incorrect price strategies, and insolvencies – any aviation incidents are subjected to immediate and painstaking analysis by the relevant national aviation authorities.

Since World War II, research into air accidents has had the main aims of identifying the causes of accidents, avoiding any recurrences, and increasing overall safety. For the accident investigators, meting out penalties or punishment has never been the primary concern. Also, since the early 1970s, the tails of commercial aircraft have each been fitted with a voice recorder and data recorder – known collectively as the "black box" – to help with accident investigations. These two devices record key flight parameters[10] such as speed, altitude, and engine performance, in addition to conversations between the pilots in the cockpit.[11] It is largely thanks to these recorders that it is now possible to reconstruct air accidents and continue improving safety in the commercial aviation industry. However, investigations into air accidents are generally extremely lengthy. By their very nature, they require laborious analysis and reconstructions.[12] Normally, investigation reports are not available until at least a year after the accident. Furthermore, they are often designated "preliminary" to indicate that other factors may also have been at play.

By the 1970s, aircraft were increasingly fitted with more reliable turbine engines. However, this did not reduce the number of accidents as much as expected. Instead, it became evident that accidents were overwhelmingly attributable to errors by the cockpit crew. At the start of the 1980s, the US Federal Aviation Administration (FAA) and NASA developed a concept to address this specific problem. Today we call it Crew Resource Management (CRM), which focuses on cooperation between the flight crew and, above all, on reducing barriers between the captain, cockpit crew, and cabin crew.

Let us turn back the clock again just to understand this "barrier" situation better. Starting in the 1930s, larger planes such as the DC-3 and Ju 52 could only be flown by a group of people working together. Depending on the aircraft, this group consisted of the pilot (captain), copilot (first officer), flight mechanic (later flight engineer, second officer), radio operator, and navigator. Generally speaking, however, there was little focus put on effective

---

10 Among other things, the flight data recorder records flight speed, altitude, flight attitude, engine performance, control instructions.
11 The cockpit voice recorder.
12 Large sections of the Boeing 747 (TWA 800) that exploded over Long Island in July 1996 were recovered from the Atlantic. The plane was painstakingly reconstructed in a special hangar in order to identify where the explosion occurred. The final investigation report was released four years after the event; cf. NTSB (2000).

cooperation between the flight crew. The now much-overused term "team" was still far in the future. At that time, the training given to pilots and the way they worked harked back to the very earliest days of flight. It was a throwback to the era of solo pilots, who, at most, had helpers to assist them in their calling.

This approach did not even change during World War II, despite the fact that bomber crews in particular had to work in teams, often in fleets of several hundred planes. For a long time, the abiding image of the pilot was of a courageous, bold, solitary hero cut from the same cloth as aviation pioneers like Charles Lindbergh and "Wild Bill" Hopson, or fighter pilots such as "Red Baron" Manfred von Richthofen. The solitary fighters who engaged the enemy in the skies while armies of anonymous soldiers battled below achieved mythical status. In reality, virtually all of these pilots flew in squadrons from as early as World War I.[13] The end of World War II and the boom in civil aviation did little to alter this traditional image. For one thing, a great many fighter pilots moved into civil aviation. Charles Elwood "Chuck" Yeager was a test pilot, a veteran of World War II, the first person to break the sound barrier in 1947, and the man celebrated in Tom Wolfe's 1979 book *The Right Stuff* as a hero, lone wolf, and fighter.

As to the cooperation within civil flight crews, it was not unknown for individual captains to exert their authority in extreme ways. One particularly impressive example features Rudolf Braunburg, an experienced fighter pilot from World War II. On joining Lufthansa as a young copilot in 1955, his American captain advised him, "Don't touch anything. This is a big aircraft!"[14]

The mythologization of the pilot was based on powerful, emotive images that transformed individuals into heroes. At the same time, these images shaped their egos and determined their authority. The restrictive influence this authority could and can produce – and the problems that arise as a result – are demonstrated throughout this book. Still, I would like to emphasize that the captains and crews described in *Confronting Mistakes* are by no means a phenomenon unique to the aviation sector. Rather, they are prototypes representing leaders in all industries. However, unlike their counterparts in the everyday corporate environment, the cockpit crews of today have learned to use a range of strategies to counteract the effects of an individual's authoritarian behavior. It was a difficult learning process for

---

13 Cf., e.g., Hackman, J.R. and Helmreich, R.L. (1987), p. 291, and Richthofen, M.F.v. (1917).
14 Braunburg, R. (1978), p. 207.

all involved, but they managed it nonetheless. Let us therefore look at cases from both before and after the implementation of the CRM concept.

The first incident dates back to 1979. It illustrates not only the absolute authority of the captain, but also the blind obedience of the copilot.

ANE flight 248: standard procedures

With more than 25,000 flying hours – almost 1,000 of them in the Twin Otter – captain George Parmenter (60) had built up a vast pool of experience. After starting his career as a pilot in the US Marine Corps, he was forced to leave at the age of 45 due to chronically high blood pressure. He then moved into the civil aviation sector.[15] Despite losing his license on a number of occasions following routine health checks, he was always reinstated after a short period and follow-up testing. As of 1970 (when he was 51), an aviation medical examiner was issuing him certificates of medical fitness on an annual basis. This meant that Parmenter was once again able to continue flying without any restrictions. In 1970, he and two partners founded Air New England (ANE), based at Logan Airport in Boston. Over the years, Parmenter became increasingly involved with managing the business and flew less and less. He notched up a mere 12 hours in the air in the three months prior to the fateful flight on June 17, 1979.

The other pilots employed by Air New England regarded Parmenter as a boss with good flying experience. However, he was also known to deviate from standard flight procedures on occasion. In at least one documented case, he flew considerably below the prescribed glide slope[16] on an approach to Hyannis-Barnstable Municipal Airport in Massachusetts. In addition, he was renowned for ignoring checklists and information from his copilots. (Pilots are expected to run through these checklists at specific intervals during the flight to ensure the plane is correctly configured and the landing gear has correctly deployed, for example.)

The copilot of ANE flight 248, Richard Roberti (32), had joined Air New England just two months earlier on a one-year contract that could be terminated at any time. With more than 4,000 flying hours, he was also

---

15 The investigation report by the NTSB provides a detailed description of how Parmenter repeatedly lost his license. Despite this, he always managed to gain a certificate of medical fitness; cf. NTSB (1980), pp. 4–6.

16 The glide slope is part of the instrument landing system (ILS). It issues an approach angle (normally 3°) to aircraft preparing to land to ensure they remain at a safe height above any obstacles until they reach the runway.

Figure 1.1   **The cockpit of a DHC-6 Twin Otter**

deemed an experienced pilot. However, he was relatively new to the Twin Otter and had not flown one until joining Air New England (Figure 1.1). ANE flight 248 took off from LaGuardia Airport in New York City en route to Hyannis at 9:32 p.m. (Figure 1.2). Both pilots had already been in the cockpit for nearly 13 hours, including waiting times. A brief break in the afternoon was the only opportunity they had had to eat a snack and drink a cup of coffee. The flight was a typical New England shuttle. During the summer months, the passengers were mainly tourists flying to and from Cape Cod, Massachusetts. On this occasion, the small two-engine Twin Otter was carrying eight passengers in addition to the two pilots.

At around 10:34 p.m., after a little over an hour in the air, copilot Roberti contacted air traffic control at Otis, an airbase belonging to the US Air National Guard that also dealt with flights approaching Barnstable Municipal Airport. The weather report issued by air traffic control put the cloud base at 200 feet,[17] visibility at three-quarters of a mile[18] in fog, wind 210° at 10 knots[19] with occasional light drizzle.[20]

The weather conditions came as no surprise to Parmenter and Roberti. They had both noticed the fog forming during the day on their earlier

17 1 foot = 0.3048 meters.
18 1 nautical mile = 1.852 km. All miles in this text are nautical miles.
19 1 knot = 1 nautical mile per hour = 1.852 km/h.
20 Three planes that landed immediately prior to ANE flight 248 reported that the cloud
    base was between 300 and 400 feet and that visibility under the cloud cover was good.

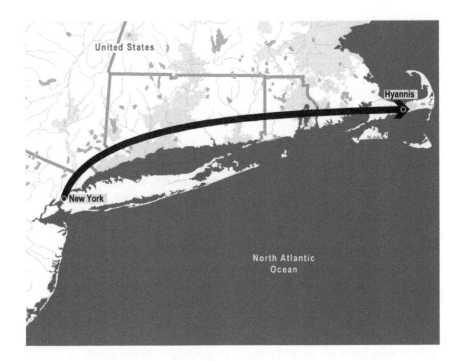

Figure 1.2 **Flight path of ANE flight 248**

flights to and from Barnstable. With the instrument landing system (ILS), landing in darkness would be tricky, but doable. Several planes had landed at Barnstable Airport just before ANE flight 248 was due, so neither Parmenter nor Roberti expected any problems.

ANE 248 was cleared to land at 10:45 p.m. Parmenter, who was flying the plane, began to descend from an altitude of 1,700 feet. The ILS required a stable approach at an angle of 3° along the electronic glide slope. At this point, the plane was slightly above the glide slope, so Parmenter began to increase the sink rate. Shortly after, copilot Roberti noticed that the ILS showed the plane was hitting the glide slope, only to drift above it once again. The plane then dropped. Roberti realized it had now entered a steep descent and was increasingly slipping below the glide slope.

As required, Roberti duly informed his captain of the drop in altitude. Furthermore, he made the required altitude call-outs as they descended, further including the decision height[21] at 200 feet. At this height, the

---

21 In aviation, altitude refers to the sea level, whereas height is the distance above ground level and used as a reference only close to the ground, i.e. landing.

runway *had* to be in sight for them to continue the approach, otherwise they would be forced to abort the approach and either come in for a second attempt or divert to the nearest alternative airport. Parmenter acknowledged none of the notifications. Roberti accepted this without further comment and the descent continued. At an altitude of 100 feet, there was still no sign of the ground, never mind the runway. Roberti said nothing and tried to catch sight of the runway.

At this point, the Twin Otter was still 1.5 miles away from the airport, behind a forested hill that gently sloped down to Barnstable. Shortly after, at 10:48 p.m., the plane brushed the tops of several tall trees and crashed into the forest. After traveling around 200 feet, it finally came to a halt with the wings torn off.

The accident took the eight passengers completely by surprise. Fortunately, the trees had absorbed some of the impact and the passengers all managed to free themselves from the wreckage. Copilot Roberti also survived the crash, although he suffered severe injuries. Captain Parmenter's body was recovered from the remains of the plane.

A medical student who happened to be among the passengers tended to those badly injured. Meanwhile, a passenger with only minor injuries tried to find her way out of the forest to get help. She eventually managed to make her way through the thick woodland to a highway, where she stopped a car. The driver took her to the airport. Barnstable Municipal Airport had already launched a search and rescue mission thanks to an off-duty Air New England pilot; having heard that ANE flight 248 was granted permission to land, he became concerned when the plane failed to appear. He set off for the approach area. There, he happened to come across some local police and informed them of the presumed crash. At 11:15 p.m., around half an hour after the crash, the state police launched a search for the missing plane, finally locating the wreckage an hour later at 12:15 a.m.

The National Transportation Safety Board (NTSB) conducted the investigation into the crash, relying on radar images and interviews with the copilot. Analysis of the radar data revealed the flight had deviated from the stipulated approach procedure.[22]

After the accident, Roberti stated that the flight parameters – particularly altitude, speed, and sink rate – had all been within the normal range at the start of the approach. However, he also said that he had noticed the

---

22 The accuracy of this data was checked using, among other things, comparisons with data from planes that had landed prior to the incident.

subsequent deviation from standard procedures as the plane rapidly sank below the stipulated glide slope. Other copilots who had flown with Parmenter confirmed that he had on occasion failed to observe standard procedure. It was therefore possible that Roberti had been in situations previously in which Parmenter had ignored the rule book. That could be why he did not step in when the same thing happened with ANE flight 248. Roberti may have assumed that captain Parmenter was planning to descend below the glide slope to break through the cloud cover and continue the approach under visibility conditions. However, even in that case, he should have realized that the rapid descent to under 200 feet could cause a crash. It was crucial that he intervene. Roberti himself told the NTSB that he had noticed the plane dropping below the decisive altitude and that he had assumed captain Parmenter had a view of the runway even though he, Roberti, was unable to see anything in the fog and darkness right up until the point of impact.

Overall, the NTSB investigation report[23] concluded that, given his general health issues, captain Parmenter may have lost consciousness prior to the accident or suffered a heart attack shortly before landing. This in itself would have been sufficient to explain why the final phase of the approach deviated from the specified flight profile. The report also identified Roberti's passive behavior as a factor contributing to the accident.

To explain this behavior, we need to look at a number of factors. In particular, we must remember that this incident occurred before implementation of the CRM concept. Quite apart from that, Parmenter was not only Roberti's captain, but also the owner of the company he worked for. If Parmenter had not died in the crash, he and his two partners would have been the ones deciding whether to extend Roberti's contract. Under such circumstances, it would be difficult to find any employee who would be prepared to alert his boss to any apparently negligent behavior or actively intervene. Roberti was also nearly 20 years younger than Parmenter. As our survey established, it is very rare indeed for younger employees to correct their superiors. However, in this instance, Roberti's internalization of the established hierarchy and his dependency led him to view the poor approach to the runway not as Parmenter's problem, but as his own. In other words, the issue was not that Parmenter was flying too low, but that he, Roberti, could not see the runway. As a result, he might have been too insecure to intervene when the plane was descending, even though

---

23 NTSB (1980), p. 21.

he must have realized how dangerous the situation had become. Clearly, restrictive authority can paralyze those lower on the hierarchy and render them unable to act even in emergencies.

Roberti may have convinced himself that an unorthodox approach to flying was just part of his captain's character. If everything had always worked out, then why not now? Even if he knew that the fog and darkness made it impossible for anyone to see the runway, he would have consoled himself with all of Parmenter's experience and felt exonerated of responsibility. In essence, we may assume that Roberti wished to avoid any conflict with Parmenter. How Parmenter would have reacted is obviously impossible to say. Yet if Roberti felt unable to make a simple remark like "I can't see the runway," we may suppose that Parmenter was a less-than-patient boss. But, who knows, he might also have corrected the problem without any objections. Either way, Roberti most likely was experiencing fear, and his silence contributed to the crash. Among other things, eliminating these kinds of issues was why the aviation industry later introduced the CRM concept.

The case of JAL flight 8054 offers a very similar scenario. In this case, once again no one intervened to prevent a series of grave errors.

JAL flight 8054: sticking your head in the sand

It is not just the yawning gaps between an older, experienced captain and a younger copilot that can cause communication problems with fatal consequences. Cultural influences also generate behavior that can hinder or completely preclude any attempt to correct errors. The following case demonstrates how a Japanese crew accepted actions and orders of a confused and disorientated American captain simply because they felt it would be inappropriate to intervene.

At 4:30 a.m.[24] on January 13, 1977, captain Hugh Marsh (53) took a cab from his motel with first officer Kunihika Akitani (31) and flight engineer Nobumasa Yokokawa (35). Their destination was Anchorage Airport in Alaska.[25] At 5:30 a.m., the trio was due to take command of a DC-8 cargo plane operated by Japan Air Lines (JAL) and fly on to Tokyo. For the cab driver who picked them up, it was a routine request. All the big airlines used Anchorage as a stopover for flights to and from Asia. This meant that planes could be refueled and the crews could be changed to comply with the legal

---

24 All times are Anchorage local time (Alaska Standard Time).
25 The following details are taken from the accident investigation report of the NTSB (1978).

requirements governing rest periods. Even in the depths of the Alaskan winter, it was not unusual for his passengers to be members of a flight crew. Nor was it unusual for Japanese crews to fly under the command of an American captain. JAL was expanding rapidly at that time and simply did not have enough experienced Japanese pilots. However, there was something different about this group of passengers. The captain made an odd impression. His face was very red, his eyes glassy, and his speech disjointed and slurred. On arrival at the airport, he had difficulty getting out of the car alone and was obviously having trouble staying on his feet. From his experience with late-night passengers, it was clear to the cab driver that the captain was drunk. When he saw that the three pilots were heading off to prepare for a flight, he decided to do something about it. He called the cab office and put in a report. Staff at the office immediately reported the incident to the JAL control room, but were unable to provide names for the crew members in question. The JAL agent who took the call promised to look into the matter and assured the caller that the airline would naturally take action if anyone was seen to be acting out of character. The agent left it at that. At 6:20 a.m., over an hour later, he finally brought the issue to the attention of the manager responsible. By that time, JAL flight 8054 was already leaving the parking position.

Captain Marsh and his two Japanese crew members knew nothing about the call from the cab office. Even if they did, it probably would not have changed anything. Marsh, Akitani, and Yokokawa boarded the DC-8 at 5:15 a.m. along with two people assigned to supervise the cargo of live cattle. The cockpit crew's subsequent preparations proceeded smoothly. Marsh clearly had no trouble running through the familiar routine, including checking the flight plan and fuel volume. Later, in the cockpit, he was able to work through the pre-flight checklists, even though his speech was still inarticulate and slow.[26] Copilot Akitani and flight engineer Yokokawa, who both witnessed Marsh's impaired motor skills during the cab journey, apparently saw no reason to question their captain's ability to fly. After nearly an hour, at 6:12 a.m., the crew was ready to start the engines. Eight minutes later, copilot Akitani informed ground control that JAL 8054 was ready to move off. The ground controller gave clearance for runway 24L. Marsh was familiar with Anchorage Airport from numerous previous flights and correctly maneuvered the DC-8 toward runways 24R and 24L, which lay parallel to one another (Figure 1.3).

At this juncture, let me take a moment to explain how takeoff and landing runways are identified by the aviation industry the world over. Runways

---

26 NTSB (1978), p. 16.

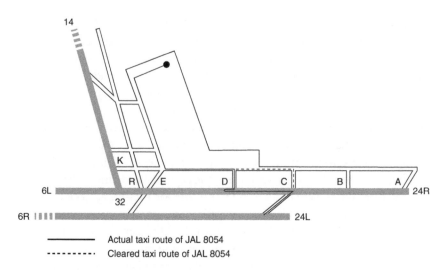

Figure 1.3 **Taxiway of JAL flight 8054 at Anchorage Airport**

are labeled using magnetic compass headings. A runway pointing 60° (east northeast) and 240° (west southwest) – depending on the direction of landing or takeoff – is identified as 6/24. If two runways lie parallel to one another, they are identified as left (L) and right (R) depending on the direction you approach it. At Anchorage, 6L and 24R are one and the same runway as are 6R and 24L. Depending on the direction in which a plane is landing or taking off, it lies at 60° east northeast and 240° west southwest. By the time JAL 8054 was cleared to taxi, planes were still landing in the opposite direction using runway 6R. While JAL began to taxi, the tower controllers were switching runway direction, as the wind was shifting from northeast to northwest and JAL 8054 received its clearance for runway 24L.

The crew prepared for takeoff as usual by running through the next checklist. While Marsh started to taxi the plane, Yokokawa read out the list of instruments and switches to be checked. Marsh and Akitani checked and confirmed these. This procedure is used in all commercial aircraft to ensure the plane is correctly configured and all the relevant systems are working properly. It was at this point that Marsh displayed the first signs of disorientation. Instead of going to taxiway C as instructed, he turned off onto taxiway D. The controller told him to halt on runway 24R before takeoff. Akitani acknowledged the instruction. The controller repeated the clearance: "Okay, Japan Air eighty fifty-four, you're going to runway two four left [24L]. Hold short of runway two four right [24R]."

The day was slightly foggy, but even in such conditions, a pilot would normally have no problem recognizing that 6L and 24R were the same runway, especially as they were marked accordingly on the ground. Feeling uncertain, Marsh turned to Akitani, "Make sure I acknowledge all transmissions. Any questions, speak up, okay?" Confused, Akitani answered, "Pardon?" "Any questions, any problems, please speak, okay?" repeated Marsh. "Yes, sir," replied Akitani. Marsh, Akitani, and Yokokawa all refocused their attention on the takeoff procedure.

Once again, the controller cleared the flight to taxi. Marsh increased the thrust while, to reaffirm the instruction, Akitani confirmed, "Clear to taxi two four." Still on taxiway D, Marsh suddenly forgot where he was and asked the controller, "What is that?" By this time, the controller had realized that JAL 8054 was unsure where to go and was not taxiing across the maneuvering area as instructed. As no other aircraft were affected, he let it go and answered dryly, "That's runway six." JAL 8054 was still on taxiway D in front of 24R.

Meanwhile, the crew was making the final preparations for takeoff, checking that the instrument displays and warning lights were working, the controls were moving freely, the trim was correctly configured, and the de-icing systems were activated. Flight engineer Yokokawa mentioned that one point on the checklist had not been covered. "Captain, yaw damper off now, okay? Usually on yaw damper."

"No," answered Marsh. Yokokawa interpreted this as an order from Marsh not to turn on the yaw damper – different than required by the specific checklist – and responded, "No okay, roger." Marsh then realized that he had indeed not turned on the yaw damper according to the standard procedures, and he rectified the action. "Not on – okay on," he said to Yokokawa. Yokokawa was relieved that the yaw damper was now turned on and responded, "Yes, sir." Marsh burst into laughter. "I've been flying eight zero one eight so long," he said. Both his Japanese colleagues laughed, even though the flight number (which was the plane's registration) was not 8018, but 8054.

Once the preparations for takeoff were completed, Akitani suggested, they should switch from the frequency of the ground control center to tower control for clearance. At this point, the plane was still on taxiway D. "Sometimes we just stay ground control, but ... " said Marsh. The frequency is usually changed once both controllers have discussed the transfer. Then he relented. "Okay. Make sure have contact." He applied thrust to make the plane taxi forward.

Akitani: "Tower, Japan Air eight zero five four, ready for takeoff."

Anchorage Tower: "Eight zero five four hold short of runway traffic landing six right [6R]."

Akitani: "Hold short, roger."

Instead of coming to a halt as instructed by the tower controller, Marsh applied more thrust and taxied onto 24R/6L. There, he stopped in the middle and set the plane up for takeoff. That he was on a wrong runway and had no clearance did not occur to him.

Akitani: "Runway six right approach outer marker. Tower said, hold short – hold short two four left."

Marsh: "Light, small airplane."

Akitani: "It's okay?"

Marsh: "Yes, sir."

Marsh: "Japan Air – holding short – we are on the runway."

Akitani [in Japanese]: "We are on the runway."

Yokokawa: "Eh?"

Akitani [in Japanese]: "We are already on the runway, yes we are."

Yokokawa [in Japanese]: "What did you say?"

Akitani [in Japanese]: "We are on the runway. Hold short, said the tower."

Yokokawa [in Japanese]: "Oh, yes, we are on the runway. This is runway two four, isn't?"

Akitani: "Two four."

Anchorage Tower: "Okay, you are on two four, right."

Marsh: "Uh – we are two four left."

Yokokawa [in Japanese]: "Heading is two four, isn't it?"

Akitani [in Japanese]: "Two four, surely."

Marsh: "Just a second."

Akitani [in Japanese]: "Even if it's a small airplane, it's a problem."

Yokokawa [in Japanese]: "Six right is in use so much before."

Akitani [in Japanese]: "It's a problem."

Marsh: "It's okay."

It is unclear from the investigation report whether Marsh actually understood Japanese. In any event, he appeared to recognize Akitani's worried tone and attempted to break off the conversation. However, Akitani realized that Marsh had taxied out onto a runway without clearance and was preparing to take off. Even so, he, too, seemed uncertain as to which of the parallel runways they were sitting on. Akitani, though, had at least noticed a plane landing on the parallel runway from the opposite direction and he reacted with due concern. In contrast, Marsh still believed he was at the start of the 10,900-foot-long runway 24L. Only the controller in the tower had a proper overview and saw that the plane was sitting on the middle of the parallel runway 24R. As this runway was not active at that time, he stayed relatively calm.

By this time it must have been more than obvious to both Akitani and Yokokawa that Marsh was not fit to fly. Akitani spoke to Yokokawa about it in Japanese. However, as the flight engineer, Yokokawa was busy with preparations for takeoff. That meant he was sitting at the console behind the pilots. He would only have been half-listening to Akitani, if at all. In any event, Akitani continued with the preparations, checked the weather minima, and ensured visibility was good enough for takeoff. Meanwhile, Marsh was making even more slip-ups.

Akitani: "Captain, takeoff minimums."

Marsh: "Takeoff minimums, okay."

Akitani: "Takeoff minimum two four left is–"

Marsh: "What?"

Akitani: "Two four left minimum is one six – sixteen hundred feet RVR [runway visual range], so it's quarter visibility fog."

Marsh: "So we have it. Thank you."

Anchorage Tower: "Okay, one-eighty on the runway or straight down to the next intersection, a right turn off this frequency, and taxi back."

Akitani [to Marsh]: "One-eighty and straight down."

Marsh: "What's our call sign?"

Akitani: "Japan Air eight zero five four, say again, please."

Anchorage Tower: "Japan Air eight zero five four, a one-eighty on the runway or straight ahead to next intersection, a right turn off the runway and taxi back down to the approach end of runway two four left and no delay in your taxi."

Akitani [to Marsh]: "Okay?"

Marsh: "Moving."

Akitani: "Eight zero five four moving."

Akitani [to Marsh]: "One-eighty and straight down to the right runway off."

Yokokawa [in Japanese]: "Going two four right – and then to the left again?"

Marsh: "Eighty five four, what can we expect?"

Anchorage Tower: "Okay, eight zero five four, straight ahead and you're approaching an intersection. For your information, you're on runway two four right. Turn right the intersection you're just approaching, it'll be about a – about a hundred-and-twenty-degree turn back to your right, then up to and hold short of two four left. You've been on runway right and should be able to get you off in just a second."

Marsh: "Okay, thank you."

Anchorage Tower: "You are welcome."

Marsh [to Akitani]: "We're cleared to two four right? Left?"

Akitani [probably points to the left]: "This side, two four left."

Akitani [in Japanese]: "We are just at the middle of two four right."

Yokokawa [in Japanese]: "Yeah, we were there."

Akitani [in Japanese]: "Made a turn from there."

In spite of the air traffic controller's directions from the tower, Marsh continued to have difficulty orienting himself. In the end, the crew nonetheless found the way to runway 24L and went through the remaining items on the checklist. At 6:33 a.m., the tower gave clearance for takeoff. At 6:34 a.m., Marsh accelerated the engine throttles for the initial takeoff thrust. The DC-8 gathered speed.

Akitani: "Power set."

Marsh: "Thank you. I have."

Akitani: "You have."

Akitani: "Eighty [knots]."

Akitani: "V one."[27]

Akitani: "Rotation."[28]

Marsh: "Rotation."

---

27 Once the plane exceeds V1 speed, the takeoff must continue, as the remaining stretch of runway no longer offers sufficient braking distance.

28 Signal that the speed is enough to lift the nose wheel. To do that, the pilot in charge pulls back the control column. This lifts the nose wheel from the ground before the plane takes off at a climb angle of between 5° and 7°. During the climb, the angle is then increased to between 10° and a maximum of 20°.

The rumbling noise of rolling down the runway gives way to the sound of being airborne.

Marsh: "Ten degrees."

Akitani: "V two."[29]

Unusual rattling and shaking noises can be heard from the plane. These continue.

Akitani: "Gear up."

Yokokawa: "Too much steep."[30]

Akitani: "Eh."

The shaking of the control column, indicating a threatening stall is clearly audible.

Yokokawa: "Stall."[31]

One and a half seconds later the sound of the impact was audible and the recording ended.

The crash just off the runway killed the three members of the cockpit crew, the two cattle handlers, and the cattle in the cargo hold.

As the authority in charge of the investigation, the NTSB was supplied with information about captain Marsh's condition very soon after the crash. However, the NTSB also examined other possible contributing factors, including load distribution, mechanical defects in the control system, the weather, and, in particular, whether ice had formed on the wings. The latter could have led to a loss of lift and the subsequent crash. However, the fact that no other plane had reported ice buildup that morning made this an unlikely cause. Analysis of the flight data recorder revealed that Marsh put the plane in too steep a climb during takeoff. He probably pulled the control column too far toward him. As a result, the plane gained altitude faster than it should have, given its weight. As Yokokawa indicated by saying "Too much steep" and "Stall," this caused the plane to stall and it began to lose altitude rapidly. Marsh probably just saw the very

---

29 The speed that ensures a safe climb even if an engine fails.

30 In original: "Too much steep." The possible alternative "Too much speed" is unlikely, given that the plane was still traveling at well below V2.

31 The plane is in a stall. The angle of attack is so large that the air flowing over the wings is no longer enough to lift the plane. The plane then loses altitude very rapidly and its control becomes restricted.

high sink rate on the vertical speed indicator, but pulled even harder on the control column. His actions sealed the fate of the entire crew.[32]

Tests of Marsh's blood eventually confirmed the suspicion that he was unfit to fly. Even 12 hours after the accident, his blood still returned values of between 298 and 310 milligram alcohol percent. According to the National Safety Committee on Alcohol and Drugs, a blood alcohol level of between 180 and 300 milligram percent causes confusion, disorientation, drowsiness, disturbed perception, poor balance, uncoordinated movements, and slurring. Marsh displayed virtually all of these symptoms prior to takeoff.

The investigation report provides no details of how Marsh usually behaved when interacting with his colleagues. For all we know, he might have been a competent, albeit slightly eccentric, pilot who behaved oddly at times. As to JAL flight 8054, however, it is impossible that copilot Akitani and flight engineer Yokokawa were not aware of the critical state Marsh was in. He must have smelled of alcohol, too.

So why did neither of them intervene? Despite the increasing concern evident in Akitani's voice, he sat helplessly by and let events run their course. As copilot, it would have been easy for him to fly the plane instead of Marsh. However, like Roberti in the previous case, the Japanese felt inhibited in the presence of his captain and clearly tried to avoid a conflict situation. The latter may also account for why he and Yokokawa failed to report Marsh to the airline, despite the obvious risks. Had they done so, it would have resulted in disciplinary action against Marsh and possibly an investigation by the aviation authority. Ultimately, Marsh could have lost his license. Furthermore, other colleagues may have construed this as a denunciation of Marsh and a violation of the comradeship within the flying community. In other words, the Japanese crew would have put themselves at the center of a furor – a prospect that certainly does not appeal to everyone. In addition, they risked being branded traitors.

In terms of their direct interaction with Marsh, both Yokokawa and Akitani were perhaps too timid or too polite to broach the issue of their American captain's intoxication. After all, a colleague's alcohol consumption is one of the many topics within organizations that are mostly discussed behind an individual's back. They might be addressed privately behind closed doors on rare occasions, but most likely never in a public forum. In addition, there can be no doubt that both Akitani and Yokokawa were extremely submissive. Just think back to the yaw damper, which simply *had* to be activated. Marsh says no, and Yokokawa – despite knowing

---

32 Given the low altitude of the DC-8 immediately after takeoff, it seems unlikely that it would have been possible to recover from the stall.

better – replies "No, okay, roger." Akitani, on his part, just submits to his captain, whether it concerns being on the wrong runway or the plane coming in the other direction – despite Marsh's instructions being muddled and confusing. In addition, the clearly drunk Marsh asked for open communication and support, which, as we shall see later, is not enough when people are unaccustomed to taking charge.

As in the previous case, we once again have a situation where a much younger colleague does not dare to question or criticize his superior. Instead, he feels obliged to accept scenarios that are fundamentally unacceptable, such as taxiing onto the wrong runway with a plane coming in the other direction. The incident with the yaw damper is a telling example of the way Yokokawa – like Akitani – manages to avoid making any direct criticism. As we saw, Yokokawa skirts around the issue by commenting, "Usually on yaw damper," instead of stating clearly "You still have to activate the yaw damper." Of course, it is possible that he told himself Marsh was experienced enough to still know what he was doing, even in a fog of alcohol. After all, Marsh had managed to work through the routine preparations for takeoff. Plus, the controllers were there to oversee the process. Like Roberti, Akitani had convinced himself that everything would be okay. In all honesty, we cannot really blame him for that. When faced with a tricky situation, passively hoping for the best rather than actively intervening is a behavior familiar to us all.

At the time, the NTSB accepted all these factors, but also concluded that Akitani and Yokokawa shared responsibility for the accident. In the investigation report, the NTSB called on the FAA to introduce measures to boost the status of crew members so that they would be more willing and able to make their views known to the captain. Still, it took a while before this recommendation started to produce results and marked another step toward Crew Resource Management, or CRM.

ALW flight 301: chaos and silence

Founded in 1988, Birgenair was an Istanbul-based charter airline. Most of its flights took passengers to the Caribbean on behalf of a major Turkish travel company. In 1995, Birgenair founded another airline, ALAS Nacionales, based in Puerto Plata, Dominican Republic. With several planes leased from Birgenair, the new airline was to fly from the Dominican Republic to various destinations in Western Europe.

On February 6, 1996, a group of 176 vacationers – most of them German – were waiting for their return flight to Germany at General Gregorio Luperón Airport in Puerto Plata. The flight took off as scheduled at 11:41 p.m. local time. Six minutes later, the plane plummeted into the Atlantic. All 176 passengers and the 13 crew members were killed.

As is often the case with plane crashes, the media initially speculated that the accident had been caused by an explosion or engine failure. Three weeks later, and with support from the US Navy, the flight data recorder and voice recorder were retrieved from a depth of more than 6,000 feet using a diving robot. Following analysis of the recorded data, it soon became clear that the accident was mainly due to inadequate cooperation between the pilots. At that time, CRM training was not mandatory for Turkish pilots.

On the evening of February 6, 1996, engineers preparing the ALAS Nacionales Boeing 767 for ALW flight 301 to Frankfurt – with two stop-overs at Gander and Berlin – discovered a fault in the hydraulic system. This made it doubtful whether the plane could be prepared in time for the planned flight.[33] However, it so happened that a Boeing 757–200 registered to Birgenair was sitting unused at the airport at this very time (Figure 1.4). The plane had been leased to an Argentinean airline until January 16, 1996.

Figure 1.4   **B757–200**

---

33 All the following details about the accident are based on the (sadly not entirely correct) translation of the unpublished English summary of the official investigation report
   (Gröning, M. and Ladkin, P. 1999; Ladkin, P. 1999a) and an edited English version of
   the investigation report (Flight Safety Foundation 1999).

After a routine engine inspection on January 23, it had been parked at the airport. No further inspections had been carried out.[34] Furthermore, the sensitive pitot tubes on the plane's exterior had not been covered, as was standard. (Among other things, these long, hollow tubes are essential for the altitude and airspeed indicators in the cockpit.) Despite these factors, the plane was chosen as the replacement aircraft for the flight to Frankfurt, flying under the ALAS Nacionales number 301. It was duly prepared for takeoff.

Although smaller than the Boeing 767, the Boeing 757 was still big enough to accommodate the 176 passengers booked on the flight. Like the Boeing 767, the replacement aircraft would be making a stop in Gander, Canada, as it was unable to complete a nonstop flight to Germany (Figure 1.5). However, the Boeing 757 was a different type of plane than the Boeing 767. That meant it could not be flown by the original crew. A brand new crew had to be scrambled at short notice. The new Birgenair crew reported to the airport at 10:15 p.m. With only one hour to prepare, the time available to them was tight but still within normal range.

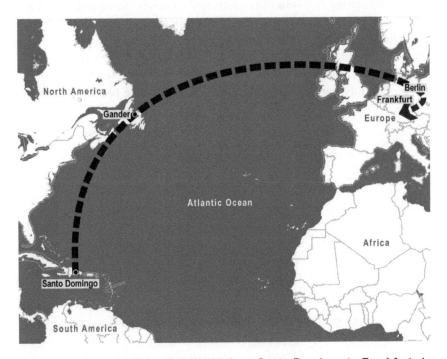

Figure 1.5 **Planned flight of ALW 301 from Santo Domingo to Frankfurt via Gander and Berlin**

34 DGAC Press Release (1996).

Still, the Boeing 757 had been out of operation for a while. This meant that the static pressure system should have been checked prior to takeoff. Among other things, this system is responsible for generating altitude and airspeed readings in the cockpit. Yet this check was not carried out, possibly because of the pressure to stick to the planned departure time. As a result, no one noticed that the left pitot tube on the captain's side was blocked, probably by an insect nest.[35] The other pitot tubes[36] were all functioning.

But to explain the fate of ALW flight 301, we need to take a closer look at the cockpit crew. The Boeing 757 was flown by the captain and copilot. Because of the length of the flight to Frankfurt with two planned stop-overs, there was also a third relief pilot in the cockpit, which is required if the flight time exceeds nine hours. He was sitting in the observer's seat behind the two pilots (Figure 1.6).[37]

The man in charge and flying the plane was captain Ahmet Erdem (62), who had built up almost 25,000 flying hours in the course of his career. He had been flying the Boeing 757 for four years, accumulating nearly 2,000 flying hours in the aircraft and thus was clearly a very experienced pilot. His copilot was 34-year-old Ayvkut Gergin. With 3,500 flying hours, he also had a fair amount of experience under his belt. However, he had only just completed his training for the Boeing 757 and had flown a mere 75 hours in this particular model. The relief pilot in the observer's seat was Muhlis Evrenesoglu (51). Like the copilot, he had only recently

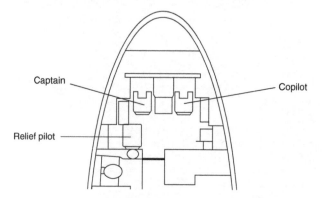

Figure 1.6 **Boeing 757 cockpit showing the seating positions of captain, copilot, and relief pilot**

---

35 Gröning, M. and Ladkin, P. (1999), p. 19.
36 The vital speed and altitude readings are based on data from three measuring devices that have redundancy built in and function independently of one another.
37 See Gröning, M. and Ladkin, P. (1999), pp. 6–8 about the crew.

completed his training for the B757 and had gained just 121 flying hours. Nonetheless, with 15,000 flying hours overall, he was also experienced.

All three pilots had completed key parts of their training at renowned international flight schools. Importantly, this was also true of the training they had each done to acquire the type rating needed to fly the Boeing 757.[38] Before taxiing off toward the runway, Erdem and Gergin once again checked the weather conditions for their departure. There were still isolated storms in the area around Puerto Plata and a light drizzle was falling from the almost solid cloud cover at 7,000 feet. The plane would take off in an easterly direction toward the Atlantic and fly into the night (Figure 1.7). At 11:36 p.m., the plane left its parking spot and began taxiing to runway 8.

The tower in Puerta Plata cleared the Birgenair flight for takeoff at 11:41 p.m.[39] Captain Erdem pushed the thrust levers forward, released the brakes, and the plane began to move off. On the captain's instructions, copilot Gergin activated the automatic thrust control system to optimize engine power before confirming that the engines had reached the pre-calculated power settings required for takeoff.

Figure 1.7   **Cockpit of a B 757–200 at night**

---

38 Pilots of large planes, particularly commercial airliners, need to hold a separate type rating for each model they fly.

39 The following description is based on the investigation report (Gröning, M. and Ladkin, P. 1999) and the transcript of the voice recorder (Flight Safety Foundation 1999).

As Gergin was not flying the plane during takeoff, he was responsible for keeping an eye on the engine indicators and other primary flight instruments. Sixteen seconds later, he reported that the plane had reached a speed of 80 knots. Following normal procedure, Erdem confirmed this information with "checked." That should have meant that his display was also showing a speed of 80 knots.[40] Yet, two seconds later, he said to his copilot, "My airspeed indicator is not working." According to regulations, there must be three working airspeed indicators on each flight (Figure 1.8). Erdem should have aborted takeoff immediately.[41] Perhaps aware of the delay this would cause, he instead asked Gergin if his indicator was working. Gergin said it was, and Erdem continued with the takeoff. After all, there were still two airspeed indicators functioning independently of one another in the cockpit. One was assigned to the copilot whereas the other, smaller one was located on the center console between the two pilots.

Figure 1.8 **Instrument panel and center console in the Boeing 757 1) captain's airspeed indicator; 2) copilot's airspeed indicator; 3) standby airspeed indicator; 4) ADI (attitude director indicator or artificial horizon)**

40 Among other things, this check is supposed to ensure that the airspeed indicator is working properly for both pilots.
41 Flight Safety Foundation (1999), p. 3.

"You tell me [the speeds]," Erdem instructed his copilot. Shortly after this, Gergin reported that the plane had reached V1. At this speed, it is no longer possible to abort takeoff. Gergin then said "rotate," indicating that the plane was ready to take off. As Erdem pulled back the control column, the nose of the Boeing 757 lifted and the plane left the ground. Gergin confirmed "positive climb." Erdem gave the order to retract the landing gear.

A little later, Erdem told Gergin to activate the lateral navigation mode of the autopilot. This displayed the autopilot's steering commands for the flight course. At this point, Gergin noticed that the captain's airspeed indicator (Figure 1.9) now matched his own and commented, "It began to operate."[42] Everything seemed to be going smoothly.

Figure 1.9   **Airspeed indicator in the Boeing 757**

---

42 In fact, the airspeed indicator was simply reacting to the pressure change in the pitot tube caused by the climb. See Gröning, M. and Ladkin, P. (1999), p. 11. Consequently, the airspeed indicator only displayed the correct speed for a brief moment. As the plane climbed higher, the speed shown was increasingly higher than the actual speed.

Once the plane reached an altitude of 1,500 feet, Erdem gave the order to switch the engines to climb mode and activate vertical navigation. He was able to monitor the speed and rate of climb using the attitude director indicator (ADI), also known as the artificial horizon. Erdem, who was still flying the plane manually, simply had to follow the pitch angle calculated by the autopilot. A short time later, the tower at Puerto Plata instructed the crew to switch their frequency to the departure controllers in Santo Domingo.

The climb was going according to plan. As the plane's speed increased, the wing flaps were retracted. Erdem and Gergin ran through the after-takeoff checklist. Gergin radioed the controllers in Santo Domingo and received permission to climb to 28,000 feet. At 3,000 feet, Erdem activated the mode L (left) of the autopilot, which would follow the programmed flight plan from that point on. Here it is important to remember that in mode L the autopilot only used data from his – in this case Erdem's – instruments.[43]

Apart from the brief failure of the captain's airspeed indicator, the flight was routine. The crew made no further mention of the earlier problem. However, unnoticed by the crew, the captain's and copilot's airspeed indicators began to show different values as the plane gained altitude. The third indicator in the center console matched the copilot's. While Erdem's airspeed indicator showed the rate of climb specified by the flight management computer,[44] those of Gergin and the standby instrument indicated the plane's speed was decreasing. The gap between the two readings continued to grow, but, at first, none of the three pilots noticed it.

At 11:44 p.m., three minutes after takeoff, the upper screen on the center console flashed the Engine Indication Crew Alerting System warnings "rudder ratio" and "mach speed trim."[45] On spotting these, Erdem read the alerts out loud and said, "There is something wrong. There are some problems." Gergin, however, was busy. He was still in radio contact with the departure controllers in Santo Domingo. Again, Erdem remarked, "Okay, there is something crazy, do you see it?" At this point, the plane

---

43 Autopilot mode R uses data from the copilot's instruments. This mode is normally activated by the copilot. For automated landings, all three modes (L, C, R) are activated and the autopilot compares all three available indicators. If one set of results deviates from the others, it aborts the approach.

44 Due to the blockage in the pitot tube, the airspeed indicator exaggerated the actual speed as the plane gained altitude.

45 This warning – which the pilots would not have been expecting at this point – refers to the position of the rudder on the tail and/or the elevator trim. It provides an initial indication that the rudder position is not consistent with the plane's speed.

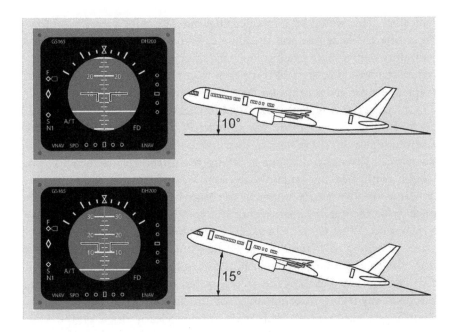

Figure 1.10   **Pitch angle and the corresponding ADI display for 10° and 15°**

was supposedly climbing at a pitch angle of 15° at a speed of 327 knots (Figure 1.10). These were contradictory values. At that pitch angle and with the engine thrust set to climb mode, the speed shown should have been far lower.

Just after, Gergin also noticed that the values displayed did not tally and commented, "There is something crazy there at this moment. Two hundred [knots], only, is mine and decreasing, Efendim."[46] Captain Erdem replied, "Both of them are wrong. What can we do? Let's check their circuit breakers."

In such situations, the normal procedure is to resolve the issue using the quick reference handbook of the aircraft. This handbook contains emergency procedures, checklists, and instructions for all imaginable faults and alerts. It is designed to ensure the flight can continue safely despite the problem. It is the job of the pilot not currently flying the plane to read out the information from the quick reference handbook. In this particular case, one of the first things the crew should have done was compare all three

---

46 The German translation of the investigation report gives this as "Effendi." This form of address is very unusual in Turkish. It is more likely that Gergin said "Efendim" (sir), indicating his respect for and trust in his captain.

indicators. This would have revealed that one of the indicators was show-ing a different value, and was therefore defective. Even if all three airspeed indicators were to have failed completely, there is a specific emergency pro-cedure that would have enabled the crew to continue flying the plane based on precisely defined settings for engine power and pitch angle.

Crucially, the Birgenair crew failed to check the quick reference hand-book. Instead, Erdem made the completely arbitrary decision that the airspeed indicator on the standby instrument panel between the two pilots was correct. Given that the value shown by this instrument matched his own, Gergin confirmed the captain's decision. He then switched his airspeed indicator to "alternate," presumably to ensure he would continue to see the values from the standby instrument. After all, the captain had decided they were correct. What Gergin obviously did not know was that in this type of plane, switching to "alternate" meant he would see the (incorrect) values from the captain's airspeed indicator. The standby airspeed indicator, in contrast, continued to show the decreasing (correct) speed. Meanwhile, the autopilot was trying to compensate for Erdem's values by increasing the climb angle. The excessive pitch angle was now clearly visible on the ADI. By this time, alarm bells should have started ringing for Erdem, if not the others.[47] Yet all he did was simply state, "As aircraft was not flying and on ground, something happening is usual, such as elevator asymmetry and other things. We don't believe them [the instruments]." Neither Gergin nor the relief pilot, Evrenesoglu, contradicted him. Evrenesoglu merely asked if he should pull the circuit breakers to reset the airspeed indica-tors. He obviously thought the problem was computer-related and hoped to resolve it by switching the indicators off and on. Captain Erdem agreed he should try it. All this time, the autopilot was flying the plane in a climb based on data from the captain's airspeed indicator. This also meant the pitch angle was constantly increasing.

The first acoustic warning went off at 11:45 p.m. The loud clacking noise indicated that the plane had exceeded the maximum permissible speed. At an altitude of 6,700 feet, the captain's airspeed indicator was reading 352 knots. Given that the pitch angle was now almost 15°, power set at normal climb mode and the plane nearly fully loaded, this speed was technically impossible. Although he knew that, Erdem said, "Okay, it's no matter. Pull

---

47 Volume 3 of the Birgenair quick reference handbook contains clear instructions for pilots flying with unreliable airspeed indicators. In particular, it lists recommended angles of attack and engine powers for a range of flight attitudes (climb, cruise, descent, and approach), depending on the gross weight. This should enable the flight to continue even without the primary airspeed indicator.

the airspeed [circuit breaker], we will see." Pulling and reinstalling the circuit breaker deactivated the loud warning alarm. Copilot Gergin asked doubtfully, "Now it is three hundred and fifty, yes?"

Neither Erdem nor Gergin bothered to look at the standby airspeed indicator in the center instrument panel. It was currently reading just under 200 knots. Still believing that the plane was going too fast, Erdem cut back the engine power to idle thrust. The pitch angle of the plane was now 18°, more than twice as steep as specified in the flight profile.

Nearly five minutes after takeoff, events suddenly snowballed. The automatic stall alarm went off, as did four other alarms. Captain Erdem must have been faced with an incredibly confusing picture. According to his indicator, the plane was going *too fast* and seemed virtually impossible to slow down. At the same time, the shaking of the control column indicated that a stall was imminent, caused by the plane going *too slow*. In addition, the autopilot had reached the limits of its steering capabilities and duly deactivated itself at a pitch angle of 21 degrees (Figure 1.11).

It was only when the plane rapidly began to lose altitude that Evrenesoglu reported the extreme pitch angle clearly visible on the attitude direction indicator, saying simply "ADI." Gergin reacted immediately and ordered, "nose down," and a moment later, "thrust." In pilot training, both these actions are among the basic maneuvers pilots are taught to regain control of a plane after a stall. At this moment, it should still have been possible to bring the plane back to a manageable attitude. Erdem ordered Gergin to deactivate the autopilot. Gergin answered, "Already disconnected, Efendim." In the meantime, the plane had dropped around 1,000 feet. Evrenesoglu said again, "ADI." Erdem, however, still seemed convinced the plane was going too fast, whereas, the plane was decelerating rapidly

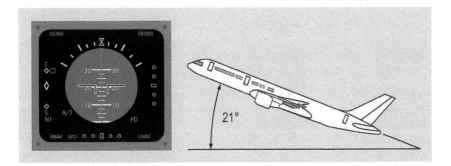

Figure 1.11 **Flight attitude and ADI indicator of ALW flight 301 at the point when the autopilot deactivated**

and dipping increasingly toward the ground. "Not climb?" asked Erdem. "What am I to do?" Even at this point, the crew could still have regained control.[48]

Gergin made an attempt to get the plane back under control using the autopilot. "You may level of, altitude okay," he said. "I am selecting the altitude hold. Altitude hold, okay, five thousand feet." Gergin had obviously forgotten that the autopilot was now deactivated. Meanwhile, the plane continued to lose altitude.

Suddenly, at 11:45 p.m., captain Erdem finally made the connection between the idle engines and the stall. He shouted, "Thrust levers, thrust, thrust, thrust, thrust!"

Clearly thinking Erdem's order was a question with regard to the position of the thrust levers, Gergin replied, "Retard."

Erdem repeated his order. "Thrust, don't pull back, don't pull back, don't pull back!"

"Okay, open, open," said Gergin and put the thrust lever to maximum power.

Although power started to build up again in the right engine, it was dropping in the left one. Suspecting Gergin was interfering in some way, Erdem called out, "Don't pull back, please, don't pull back." Gergin responded that he had set the engines to maximum thrust. "Open, Efendim, open."

Due to the asymmetric power in the two engines, the already unstable plane now rolled to the left until the wings were standing perpendicular to the ground at an angle of almost 99°. The nose of the plane was pointing steeply downward (−58°).

Evrenesoglu: "Efendim, pull up."

Erdem: "What's happening?"

Gergin: "Oh, what's happening?"

Then the ground proximity warning system (GPWS) sounded.

Gergin's last words were "Let's do like this."

Fifteen miles northeast of Puerto Plata, the plane smashed into the Atlantic. All 176 passengers and the 13 crew members were killed instantly.

---

48 This was demonstrated as part of tests in a flight simulator. Furthermore, Boeing reported that, during a test flight, a B757 flying at a comparable altitude had gone into an unplanned stall. In this case, the crew had been able to regain control using standard maneuvers (nose down, rudder and ailerons neutral, increase power). Cf. Flight Safety Foundation (1999), p. 6.

As the plane had not issued an emergency call, it was some time before it was reported missing. Hours after the crash, ships discovered scattered pieces of wreckage floating on the surface of the ocean. Following a request from the Dominican government, the US Navy later provided special equipment for retrieving wreckage from deep underwater. The Federal Republic of Germany and the companies Boeing and Rolls Royce also provided financial assistance to support the search for wreckage and the flight data recorder. The data recorder and voice recorder were recovered from a depth of 7,200 feet 22 days after the crash on February 28.[49] The investigation could begin.

Nine months after the accident, the official incident report[50] was published by the Dominican Republic's Dirección General de Aeronautica Civil, supported by the US National Transportation Safety Board. The information contained on the flight data and voice recorders revealed that the airspeed indicator on the captain's console had been faulty.[51]

The investigation resulted in a number of conclusions.[52] The probable cause of the crash was the crew's failure to correctly interpret the stall warning. As a result, they were unable to take the necessary steps to regain control of the plane. The crew had become confused due to the contradictory readings on their indicators and the way events accelerated. Their flying skills also proved inadequate in this crisis situation and they showed limited knowledge of this particular type of plane. In addition, the investigation highlighted the poor quality of the maintenance work carried out on the plane during its time on the ground, including the failure to cover the pitot tubes. Furthermore, the static pressure system had not been checked before the plane was put into operation again. The engineers on the ground should have ensured both these things were done. However, the key factor behind the crash was the lack of proper cooperation between the crew members.

It may become easier to understand Erdem's behavior – which was largely responsible for the crash – if we remember that he was an experienced pilot. For him, this flight posed no particular challenge. He was not even worried by the first alerts. After all, he was familiar with the redundancy built into

---

49 Ladkin, P. (1999b).
50 Extracts translated into German and English; cf. Gröning, M. and Ladkin, P. (1999) and Flight Safety Foundation (1999).
51 It was not possible to identify the exact cause of the blockage in the pitot tube, which was never found after the crash. The blockage had caused the faulty readings on the airspeed indicator and was probably due to an insect nest or dirt particles. Cf. Flight Safety Foundation (1999), p. 7.
52 See Gröning, M. and Ladkin, P. (1999), pp. 21–22.

the systems of the Boeing 757, and the flight appeared to be progressing as expected. When the first system errors were triggered as the plane began its climb, the aircraft nevertheless seemed to be flying normally. Erdem thus saw no reason to worry and slipped into a mindset that is common to us all. By telling himself everything was "normal," he created the framework of explanations for everything that was to follow. As a result, he saw what he *expected* to see: namely, a scenario familiar to him from his many years of experience. All of us have experienced this form of cognitive distortion. Given that it happens subconsciously, it probably happens more often than we think. Initially, even the contradictory information could not disrupt Erdem's fixed mental image. He only became concerned when what he saw no longer fit within his set of expectations. But even then any feelings of uncertainty were absorbed by his mental attitude and anything not normal termed "crazy" and "wrong" and unbelievable. "As aircraft was not flying and on ground, something happening is usual [...] We don't believe them [the instruments]."

As tragically as this flight ended, Erdem's case offers just one example of how we base experiences on what has happened to us in the past and take this as rock-solid knowledge. As such, it serves as a necessary reference system, but it can never be trusted as an *infallible* basis for making sound decisions now or in the future. For Erdem, it was what he thought he knew that hampered his ability to actually think. Consequently, he continued to believe in his concept of "normality" even when the stall warning indicated that they were in an emergency situation.

But the fact that Erdem was mistaken and persisted in his mistaken beliefs was just the first act in this particular tragedy. Gergin also played a key role in events. Once again, we have a second, younger man who subordinates his views and opinions to those of his superior officer. Physically, the contradictory indicators were located fairly close to one another in the cockpit. If the two men had been of equal status, they might have operated on a "check and balance" system; Gergin would have compared the captain's indicator with the other two and raised the matter of the contradictory values. As it was, he quietly accepted Erdem's assertion that everything was fine. If there was a problem, it was due to a computer error, not an error by the captain. Consequently, the computer was effectively sidelined, that is, restarted to deactivate its alerts.

Just like Roberti, Yokokawa, and Akitani, Gergin is passive. It is true that he hastily makes the routine reaction of "nose down" straight from the training manual when the danger becomes overwhelming. Otherwise, he is as helpless as his counterparts in our first two cases. Of course, the confidence and certainty exuded by another person can have a compelling influence on us, especially, when that person apparently has simple and plausible explanations for difficult issues. This situation may be compounded further by the authority and decision-making power of an "Efendim," whose inferiors have not learned to criticize him or would not dare to actually try. Without doubt these hierarchical distances have more than once resulted in scenarios like those in the cockpit of our Birgenair flight, even if the consequences have been less tragic.

KLM flight 4805: in a hurry

Air accidents have been occurring ever since the Wright brothers made the very first powered flight in 1903. Despite this, state-of-the-art planes and experienced, highly qualified crews have always been seen as the key to safety in the air. If this thesis had not been in doubt before, it was rocked to its core on March 27, 1977. On that day, two Boeing 747s collided on the runway on the island of Tenerife and burst into flames, killing 583 people. It remains the worst disaster in the history of civil aviation. The accident occurred after a Boeing 747 belonging to Dutch airline KLM began its takeoff before a Pan Am Boeing 747 had time to leave the runway. Both planes were in the hands of extremely experienced crews. In the beginning, it was impossible to grasp how this dreadful accident could have happened.

For captain Jacob Veldhuyzen van Zanten (50), KLM flight 4805 from Amsterdam to Las Palmas on Gran Canaria was a welcome change from his management responsibilities.[53] As director of the KLM Flight Training Department, he had spent the past six years in charge of training and testing KLM pilots. As a result, he only flew normal scheduled flights if there was a staff shortage.

For the two other members of the crew, it was a flight like any other. With 9,200 flying hours under his belt, first officer Klaas Meurs (32) was nearly as experienced as captain van Zanten, despite being 18 years younger. The only thing that made the flight slightly out of the ordinary for Meurs was the fact that, two months previously, van Zanten had been his check pilot for his type rating for the Boeing 747. For flight engineer Willem Schreuder (48), who had already completed 15,000 flying hours, it was a completely routine assignment.

---

53 All the following information on the accident is based on the Aircraft Accident Report drawn up by the Air Line Pilot Association, the accident report of the Comisión de Investigación de Accidentes e Incidentes de Aviación Civil and the investigation report by the Netherlands Aviation Safety Board. Roitsch, P.A. et al. (1979), Comisión de Investigación de Accidentes e Incidentes de Aviación Civil (1979), Netherlands Aviation Safety Board (1977).

The events leading to the tragedy began when KLM 4805 headed from Amsterdam to Gran Canaria with 234 passengers on board. While they were over the Atlantic, the police in Las Palmas received a worrying call just after 1 p.m. local time. The call warned of a bomb planted in the airport by Canarian separatists. As a result, Gando Airport in Las Palmas was closed and the people inside were ordered to evacuate. At 1:15 p.m., a small bomb exploded in the airport's only terminal, injuring eight people. With fears of further explosions, the airport remained closed and all planes that were due to land there were forced to divert to other airports.

For KLM 4805, that meant changing course and heading toward the much smaller Los Rodeos Airport on Tenerife. The flight landed there at 2:38 p.m. Due to the lack of space, it and many other planes had to park on a taxiway rather than pull into assigned spaces on the apron. Forty minutes later, another Boeing 747 arrived – Pan Am flight 1736, which had been en route from New York to Las Palmas when it was also diverted to Tenerife. The crews of both planes hoped that Las Palmas would open again soon so they could land there and offload their passengers.

The uncertainty about when they might be able to leave Tenerife was a particularly unwelcome development for captain van Zanten. Around three months before, the Dutch aviation authority had introduced new stricter flight time limits for pilots. Any pilot exceeding the maximum permitted flight time risked a fine, loss of license, or even prison. Aware of these restrictions, van Zanten started consulting with KLM's flight planning department in Amsterdam after landing. If the crew was to stay within the permitted time limits, the plane would have to start its return flight from Las Palmas to Amsterdam no later than 8 p.m. local time. Otherwise, it would be impossible to embark on the return flight and the airline would have to organize hotel rooms for more than two hundred passengers in Las Palmas.

Although the airport on Gran Canaria was expected to reopen soon, van Zanten was worried that all the diverted flights would cause significant delays. With that in mind, he decided to instruct the passengers to disembark so that he could refuel the plane at Los Rodeos for the return flight to Amsterdam. This would save time later in Las Palmas. Just as the KLM Boeing 747 started refueling at 3:30 p.m., Las Palmas reopened. Unlike most of the other planes, KLM 4805 was now stuck where it was. It took until 5:45 p.m. to finish refueling and get all the passengers back on board.

Now, however, there was another problem. The weather conditions had deteriorated and low clouds were moving in over the high plateau where Los Rodeos Airport stood, creating a cloak of fog on the airfield. Van Zanten became impatient and urged his crew members to hurry. "Hurry, or else it will close again completely."

At 5:56 p.m., KLM flight 4805 in Los Rodeos was cleared to taxi (Figure 1.12). As large sections of the taxiways were still blocked by parked planes, KLM 4805 was told to taxi over runway 30 to reach its position on the other side of the airfield. Shortly after, Pan Am 1736 was also cleared to taxi via the same runway, following the KLM flight. With the fog now much thicker, the Pan Am crew soon lost sight of the KLM plane in front.

Both planes were supposed to leave the runway via the third taxiway (C3) and follow the usual taxiway, B, to their position on runway 30. However, both the KLM and Pan Am crews had difficulty finding C3. Bear in mind that the pilots in a Boeing 747 are nearly 28 feet above the ground. At a small airport with limited signage, and in thick fog, it is hardly surprising that the crews found it hard to spot the taxiway.

In response to a question from the tower, the KLM flight reported that it had just passed the final taxiway (C4). The tower instructed the crew to continuing taxiing to the end of the runway. There, they were to turn the plane 180° and wait until the Pan Am flight had left the runway via taxiway C3.

At 6:03 p.m., KLM 4805 was standing ready on runway 30. Meanwhile, the Pan Am flight was still creeping through the thick fog. Unlike the KLM flight, the Pan Am crew had managed to spot taxiway C3. However, due to the narrow curve radius, it was almost impossible to make the turn with a Boeing 747. The crew decided to taxi forward to C4, where the curve radius was more suitable for such a big aircraft.

In the KLM plane, captain van Zanten was waiting impatiently for clearance to take off. Due to his decision to refuel, the onward flight had been delayed further. It was going to be even more difficult to stick to the latest possible takeoff time of 8 p.m. for the flight from Las Palmas to Gran Canaria: on arrival at Gando Airport in Las Palmas, the passengers from Amsterdam would disembark; then the plane would be cleaned as planned before a new set of passengers boarded for the return flight to Amsterdam; and meals for the flight had to be brought on board. Every additional delay

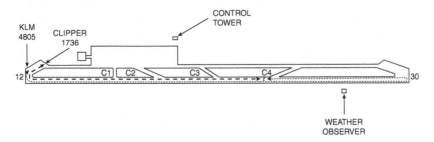

Figure 1.12 **Los Rodeos Airport and runway 12/30**

in Tenerife was shaving precious minutes off the available turnaround time. For a pilot used to scheduled flights, this was not an unusual situation to be in. Particularly during the winter months, delays or even cancellations due to poor weather were part and parcel of the job. However, for the past six years, van Zanten had only flown such flights in exceptional cases and was no longer used to such disruptions. As a check pilot,[54] he also spent a lot of time in flight simulators. Consequently, he was used to *creating* situations rather than finding himself stuck in them.

At 6:04 p.m., van Zanten instructed first officer Meurs to run through the pre-flight checklist. Immediately after Meurs completed the checklist, van Zanten began to push the thrust levers and prepare the engines for takeoff. He had given no further instructions to the crew, nor had the controllers in the tower cleared the flight for takeoff. Meurs intervened and said, "Wait a minute, we do not have an ATC clearance." Van Zanten replied impatiently, "No, I know that, go ahead, ask."

Meurs contacted the tower to report that the plane was ready for takeoff and to request clearance. The tower duly responded, "KLM eight seven zero five [sic], you are cleared to the Papa beacon. Climb to and maintain flight level nine zero. Turn right after takeoff. Proceed with heading zero four zero until intercepting the three two five radial from Las Palmas VOR."[55] This gave van Zanten his instructions for the flight path after takeoff but, crucially, did not issue a takeoff clearance. Nevertheless, he uttered a relieved "Yes."

While Meurs repeated the information over the radio, van Zanten applied thrust once again. Even if Meurs did not see what the captain was doing, he would have heard the tone of the engines changing. However, this time he said nothing. Meurs radioed the tower. "We are now – taking off." This message could have meant either that the KLM flight was in the takeoff position or that the plane was already on the move. After initially acknowledging this message with an "okay," the controller in the tower then added, "Stand by for takeoff. I will call you." Unfortunately, the second part of this message was drowned out by a radio call from the Pan Am crew, who were worried that the KLM flight might try to take off while they were still taxiing on the runway. If two messages are sent simultaneously on the same frequency, then the signals become superimposed, creating shrill static noise. This static can only be heard by third parties and

---

54 The official title is Type Rating Examiner.
55 VOR: Very High Frequency Omnidirectional Radio Range – a radio range used for navigation.

generally makes it impossible to understand any of the messages transmitted. Crucially, those directly involved are unaware that their transmissions have been affected. Consequently, the KLM crew only heard the "okay" from the controller, followed by an incomprehensible whistling noise. For their part, the tower controller and the Pan Am crew would have assumed their messages had been heard and understood. The frequency now became free again. The controller asked the Pan Am crew to confirm if they had left the runway. Up till now, he had addressed the crew with "Clipper," the normal call sign for Pan Am flights. On this last call, however, he used the more unusual "Papa Alpha."[56] Although the Pan Am crew would have understood it, it is possible that the KLM crew did not.

Due to the fog, neither of the two crews could see that their planes were on a collision course. Even so, flight engineer Schreuder was concerned enough to ask, "Is he not clear then?" Busy trying to keep the accelerating Boeing 747 in the center of the runway in foggy conditions, van Zanten did not catch what he said. "What did you say?" he asked. "Is he not clear – that Pan American?" repeated Schreuder. "Oh, yes," said van Zanten and Meurs.[57]

The Pan Am crew was now worried by the radio transmission from the KLM plane saying, "We are now – taking off." Captain Victor Grubbs growled, "Let's get the f – – out of here." First officer Robert Bragg added, "Yeah, he's anxious, isn't he?" Flight engineer George Warns also chipped in, "Yeah, after he held us up for an hour and half, that f – -." At that moment, captain Grubbs spotted the outline of the KLM plane racing toward them and shouted, "There he is – look at him – f – – that – that f – – is coming!" Bragg cried, "Get off! Get off! Get off!" In an effort to get his plane off the runway, Grubb gave full power on both right engines.

Meanwhile, in the cockpit of the KLM plane, Meurs reported that they had reached V1.[58] That meant the takeoff could no longer be aborted. Four seconds later, van Zanten saw the Pan Am plane, which had just managed to get its front wheels on to the grass. "Oh, f – – ," was all he said. His plane was still not going fast enough to take off. Nonetheless, van Zanten attempted to fly up and over the Pan Am plane. KLM 4805 did manage

---

56 The flight number of the Pan Am plane was PA 1736. In radio transmissions, however, Pan Am flights were always addressed with the call sign "Clipper" (in this case "Clipper 1736").

57 The cockpit voice recorder transcripts deviate with regard to this quote. While the Spanish accident report and the Air Line Pilot Association Report mentioned only van Zanten speaking, the Dutch report also showed Meurs saying "Yes."

58 Once V1 is reached, the remaining stretch of runway is no longer long enough to halt the takeoff, and there is no choice but to proceed.

to lift off the ground before reaching the Pan Am flight, but it was too little, too late. Part of its landing gear caught the rear of the Pan Am cabin, while the engine on the far right ripped off the roof of the Pan Am cockpit. The KLM plane flew for about another 500 feet before crashing into the runway and bursting into flames. All 234 passengers and 14 crew members were killed instantly. The Pan Am plane was also in flames – 317 passengers and nine crew members died. Only 63 Pan Am passengers in the front section of the plane and seven crew members survived the disaster.[59]

Due to the fog, the two controllers in the tower initially only saw an explosion. Then they began to receive radio messages from other flight crews and realized something terrible had happened. The airport fire service reached the burning wreckage of the KLM plane first and tried to extinguish the fire. At first, because of the fog, they failed to see the burning Pan Am plane lying 500 feet away. The full extent of the disaster only became apparent when the fire crew started seeing people who had escaped the Pan Am wreckage (Figure 1.13).

Figure 1.13   **Wreckage of the KLM 747**

---

59 A further nine passengers died later due to the severity of their injuries.

The accident on Tenerife was the subject of a detailed investigation by the Spanish, Dutch, and US aviation authorities. The catastrophic accident was caused by a number of factors. One of them was that the Spanish controllers' radio transmissions deviated from standard procedure. However, there was *one* individual who played a central role in the fatal events: captain van Zanten. He made the disastrous decision to go for takeoff without waiting for clearance from the tower controller.

During the investigation, it naturally emerged that the captain of the KLM flight was no ordinary pilot, but the head of flight training and safety. Aviation experts were left dumbfounded by this revelation. Following the analysis into the cause of the crash, the investigators also summarized the factors that contributed to van Zanten's poor decision.[60] It became clear that, although he was in charge of pilot training, it had been a long time since he himself had had any practical experience with scheduled flights. He must have found the bomb explosion at Las Palmas and the subsequent disruption to his flight extremely stressful. Plus, he had become unfamiliar with the routine involved in scheduled flights. It was obvious that the pressures caused by the restrictions on flying time made him increasingly impatient. As we all know, rational thinking often goes out the window when an emotion or feeling takes hold. In these situations, mistakes can happen. In this case, what was no more than clearance for departure was hastily misinterpreted as clearance for takeoff. However, one question remains: why did the other two members of the KLM crew not intervene?

Let us take a closer look at the cockpit team. The KLM plane was in the hands of a hierarchically structured team, with captain van Zanten at the head. His authority was even more absolute given his status as the man in charge of flight training. Although he was not responsible for disciplining other pilots, he did oversee their stipulated six-monthly check flights. These flights determined whether a pilot's license was extended. Van Zanten was also responsible for training and issuing licenses to new pilots. In this role, he had supervised Meurs' check flight for his license to fly the Boeing 747. Even so, the KLM cockpit recordings indicated that the atmosphere was very focused, but by no means overly tense. Like van Zanten, Meurs was also well aware of the time pressure. Both of them wanted to get off the ground as soon as possible. When, after turning the plane, van Zanten applied thrust before clearance was granted, Meurs tried to intervene and said, "Wait a minute, we do not have an ATC clearance." Clearly irritated,

---

60 Weick provides an excellent summary of the investigation reports; cf. Weick, K.E. (1990).

van Zanten snapped: "No, I know that, go ahead, ask." This may be when the mood in the cockpit changed. The next time, when van Zanten applied thrust after receiving departure instructions but no clearance for takeoff, Meurs did not object. He simply contacted the tower to report "We are now – taking off." Maybe he hoped the tower controller would stop them. In truth, the situation remained unresolved. Van Zanten saw things differently. For him, everything was crystal clear. His plane would shortly be in the air and he would be on his way to Gran Canaria, then Amsterdam. If he expected his crew to dispute his actions, their silence would have confirmed his belief that he was right.

Unlike the previous cases, we are dealing here not just with the authority of a superior, but also his *mood,* which set the tone for the rest of the crew. As we saw in the Birgenair incident, the mindset of the captain can become the be-all and end-all. In van Zanten's case it was his emotional state that dominated events and influenced the crew. Van Zanten was impatient, and the other two were probably keen not to rile him further, but wanted to prevent his impatience from escalating and turning to anger.

Once again, we can empathize with the way our protagonists behaved. After all, such scenarios are familiar to all of us. In fact, we start internalizing these behaviors as children. We all had parents or teachers whose moods and feelings made it clear to us whether they were pleased or annoyed with us. If they were pleased, we learned to be happy and grateful. If they were annoyed, we felt ill at ease. Equally, we discovered how we could induce happiness and avoid generating anger. Interestingly enough, we rarely seem to be able to detach ourselves from this causality. In later life, we unquestioningly seem to transfer it to our roles as managers and employees. As a result, a manager's mood dictates the atmosphere among his employees, even if he does not consciously intend it to. By the way, it is not just negative feelings that can influence others – a visibly good mood can have the same effect. Just think of the sense of euphoria generated by a new project with potential for success. In this situation, what employee would point out problems and thus rain on the parade, particularly if the rest of the staff is equally caught up in the general high?

In all the examples we have looked at, the tendencies to submit to authority, worry about possible reactions, or fear what will happen if the authority figure gets annoyed ultimately had fatal consequences. All that the copilots had to do was ask, give a warning, issue a correction, or – as we saw in the last case – insist on compliance with a specific regulation. This would have reminded van Zanten that he was not allowed to take off without explicit permission from the tower. It is possible that criticism would not even have annoyed the captains in our case studies. Even if it had, the crews would hardly have found themselves in situations any worse than the accidents that subsequently unfolded. Yet the fact remains that they did not intervene, as they lacked the confidence to correct someone higher up the hierarchy. It was only later, with the advent of the CRM model that *safety, checking, and double checking* became the highest authorities in any cockpit communications.

UAL flight 173: I told him so

On the afternoon of December 28, 1978, an eight-member-strong United Airlines crew consisting of two pilots, a flight engineer, and five flight attendants was preparing United Airlines 173 (UAL 173) for its flight from Denver to Portland, Oregon[61] (Figure 1.14).

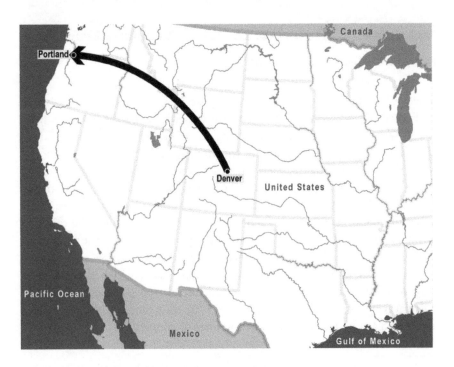

Figure 1.14 **Route of UAL 173 from Denver to Portland**

---

61 The following facts are based on the report issued by the National Transportation Safety Board, NTSB (1979).

Figure 1.15   **DC-8–61 during takeoff**

The McDonnell-Douglas DC-8–61 (Figure 1.15) was almost full when it embarked on its journey in the early afternoon, only slightly behind schedule. It was carrying 181 passengers, including six children and an off-duty pilot sitting behind the captain on the jump seat.[62] Three and a half hours later, the wreckage of the DC-8 lay in a residential area six miles south of Portland International Airport. Eight passengers and two crew members were killed. A further 21 passengers and two crew members suffered severe injuries. There was a comprehensive investigation into the cause of the crash. It revealed that, even though the plane was under the command of one of the airline's most experienced pilots, this accident was once again attributable to human error. Unlike previous cases, which the US aviation authorities had regarded as isolated incidents, a pattern now started to emerge. This showed that the planes' increasing technical reliability had not in fact led to any decline in the number of accidents.[63]

---

62 The NTSB investigation report has no further details about this captain. In particular, there is no information about what types of planes he was licensed to fly.

63 An analysis by Boeing showed that over two-thirds of all air accidents were the result of human error. Some of the particularly spectacular accidents prior to 1979 that confirm this trend include the crash of an Eastern Airlines L-1011 over the Everglades and the previously mentioned collision between two jumbo jets on the runway on Tenerife.

At the end of the 1970s, United Airlines was the largest airline in the Western world. Serving mainly flight routes within the United States, the company enjoyed an excellent reputation. United had a first-class technical maintenance service, and fulfilled the very highest standards, particularly when selecting pilots.

The captain of UAL 173, Malburn McBroom (52), had been flying for 27 years. Of those, he had spent 19 years as a captain with United Airlines. With more than 27,000 flying hours, he was one of United's most experienced pilots. He had been flying the DC-8 for several years. The other two members of the cockpit crew were also experienced pilots. The 45-year-old copilot and first officer, Roderick Beebe, had been flying for United for 14 years, but had only spent four months piloting the DC-8. The flight engineer and second officer, Forrest Mendenhall (41), had been flying with United for 11 years and was just as familiar with the DC-8 as the captain.

Although flight crews fly in different teams due to shift rotations, the cockpit crew of UAL flight 173 had been working together for several days and knew each other relatively well. The five-member-strong cabin crew was led by purser Joan Wheeler (36).

The DC-8–61 had been operated by United Airlines for slightly more than 10 years. During that time, it had notched up a total flying time of 33,114 hours. It was not one of the airline's most modern planes, but it was the typical, reliable workhorse of that era. The period it had been in operation was normal even by today's standards.

In the early afternoon of December 28, the cockpit crew met as usual around 90 minutes before UAL 173 was due to depart from Stapleton International Airport in Denver. The crew would take over the controls of the incoming DC-8 from John F. Kennedy International Airport (JFK) in New York and continue the flight to Portland. According to the weather forecast for Portland, it was going to be a clear evening with visibility of more than 15 miles, a temperature of -2° C, and a moderate northerly wind. The crew did not expect any serious delays. Captain McBroom duly ordered 46,700 pounds of fuel. This amount was sufficient for the expected 2.5 hour flight time, the 45 minute extra flight time required by law, as well as an additional 20 minute buffer, according to United's company policy.[64] McBroom and copilot Beebe agreed that Beebe would fly the plane while McBroom took charge of radio communications.[65] As the third member of the cockpit crew, flight engineer

---

64 NTSB (1979), p. 24.
65 It is normal practice for the captain and copilot to take it in turns to fly the plane. However, overall responsibility always remains with the captain.

Figure 1.16 **Cockpit with captain, copilot, and flight engineer**

Mendenhall was responsible for operating and monitoring the plane's technical systems. He sat behind the two pilots in the cockpit (Figure 1.16).

The cockpit crew then discussed details of the flight with the cabin
crew. Like the cockpit crew, the flight attendants and Wheeler had agreed
on their respective roles within the cabin. Once the briefing was complete,
the crew made their way to the DC-8 arriving from New York. Due to their
clearly defined roles, each member of the crew knew exactly what his or her
tasks were.

In this post-Christmas period, UAL flight 173 was almost fully booked.
At 2:33 p.m. Pacific Standard Time,[66] the plane was pushed back from the
gate in Denver and taxied to the runway. Fourteen minutes later, copilot
Beebe got the DC-8 off the ground and set course for Portland. After a
routine flight, captain McBroom contacted Portland air traffic control at
5:05 p.m. before the plane was cleared for visual approach to runway 28
left. The flight attendants gathered up the last remaining glasses and prepared the cabin for landing. Beebe already had a good view of the runway.
Five minutes later, he set the flaps for landing and gave captain McBroom
the order to extend the landing gear.

---

66 Time according to the place of the later accident.

During this process, the crew and the passengers all heard a loud noise and felt a powerful jolt. Captain McBroom assumed that the landing gear had deployed more quickly than usual. He also noticed that, of the three landing gear indicators, only the green light for the nose landing gear showed it was down and locked. Due to the flow noise, the pilots knew that part of the landing gear was deployed. However, they did not know if the main landing gear under the wings was fully extended and properly secured. McBroom reported that the plane was pulling to the right, indicating an asymmetrical configuration. As the air traffic controller in Portland instructed UAL 173 to contact the tower in preparation for landing, McBroom answered, "Negative, we'll stay with you. We'll stay at five [thousand feet]. We'll maintain about a hundred and seventy knots. We got a gear problem. We'll let you know." The controller told the crew to stay at 5,000 feet and set a course of 220°. "I'll just orbit you out there 'til you get your problem [solved]."

As the plane still had enough fuel to fly for more than an hour, the crew focused on the problem with the landing gear. If it was deployed only partially or not at all, there was a danger that one of the wings might touch the ground during landing. That would make it impossible to control the plane. Consequently, the cabin crew would have to prepare the passengers for a possible emergency landing and evacuation.

At 5:14 p.m. (the time at which the plane was scheduled to land in Portland), the controller put the plane into a holding pattern southeast of Portland. While the cockpit crew went through the checklists for landing gear faults, purser Wheeler entered the cockpit. Captain McBroom explained the problem to her. He said that he would tell her what he planned to do next, once he had carried out a few more checks. Meanwhile, copilot Beebe went into the cabin and tried to see the visual indicators on the wings. If these were extended above the wing surface, it would show that the main landing gear had been extended and secured. However, it was already too dark to tell.

At 5:38 p.m., around 23 minutes after entering the holding pattern, McBroom contacted the United Airlines maintenance center in Portland to explain the problem. When the flight service advisor asked whether McBroom could confirm an estimated landing time of 6:05 p.m., McBroom replied, "Yeah, that's good ball park. I am not going to hurry the girls. We got about a hundred-sixty-five people on board and we want to take our time and get everybody ready and then we'll go. It's clear as a bell and no problem."[67]

---

67 NTSB (1979), p. 3.

Let us take a closer look here at the people who make up the cabin crew. Operational responsibility lies with the purser. The purser and his or her team of flight assistants take care of the passengers and, in an emergency, are responsible for the emergency exits and evacuating the plane as quickly as possible. Like the cockpit crew, they receive regular training to ensure they can cope with emergency situations.

As the cockpit crew has basically no contact with the cabin, they rely on the flight attendants to pass on relevant cabin information. In the 1970s, the hierarchical structures between the cockpit and cabin crews meant communications were more or less always between the purser and the cockpit.

On the day of UAL flight 173, the problem with the landing gear did not initially appear to be a major issue. Between 5:14 and 5:45 p.m., the captain and copilot switched roles. McBroom was now flying the plane while Beebe operated the radio. In addition, McBroom was still talking with United's maintenance center.

At 5:45 p.m., purser Wheeler came into the cockpit again. She reported that the cabin crew was ready to announce the planned emergency landing. Shortly after, Beebe asked flight engineer Mendenhall how much fuel was left. He replied that there were still 5,000 pounds.

McBroom probably failed to hear this update as he was discussing the emergency landing with the pilot flying with them behind him on the jump seat.[68] The off-duty pilot then went into the cabin to help the flight attendants. Two minutes later, Beebe asked for another fuel update. This time McBroom replied that there were 5,000 pounds left and pointed to the flashing warning lights that would show if the remaining fuel fell below 5,000 pounds. In other words, it should have been clear to everyone in the cockpit at this point that there was enough fuel for another 20 minutes; together, the four engines consumed around 220 pounds per minute.[69] A little later, McBroom asked the flight engineer to calculate the landing weight – based on the zero fuel weight[70] of the plane and the remaining fuel – in "about another fifteen minutes." Mendenhall repeated, "Fifteen minutes?" Brushing this off, McBroom simply added that Mendenhall should factor in three or four thousand pounds on top of the zero fuel weight. It is worth taking a closer look at that calculation. Together, the four engines would consume 3,300 pounds in the space of 15 minutes. If the plane were to land in 15 minutes' time, the

68 He was only flying in the cockpit as a passenger.
69 NTSB (1979), pp. 18–19.
70 Empty weight of the plane plus passengers and cargo.

amount of fuel remaining would be 1,700 pounds, certainly not 3,000–4,000 pounds.

"Not enough," murmured Mendenhall. Raising his voice a little, he said, "Fifteen minutes is gonna really run us low on fuel here." He made it no clearer than that. Shortly after, he notified the captain of the plane's landing weight. Still, McBroom was assuming a fuel reserve of 3,000–4,000 pounds while Mendenhall was working with a reserve of just 1,700.

By now, the plane was 18 miles south of Portland Airport and flying in a curve toward runway 28. The United flight service advisor requested the plane's estimated landing time.

At 5:53 p.m., Mendenhall asked McBroom if they would land at "about five after" [i.e., 6:05 p.m.]. McBroom said yes. Three minutes later, Beebe asked Mendenhall about the fuel situation. "Four, four – thousand – in each [tank] – pounds," Mendenhall replied vaguely. "Okay," said Beebe, even though Mendenhall's answer could have meant 4,000 pounds overall or 4,000 in each tank. Naturally, neither Beebe nor McBroom would have thought there were still 16,000 pounds left at this point. For that reason, let us assume that they were basing their calculations on an overall fuel reserve of 4,000 pounds.

McBroom said to Mendenhall, "Walk back through the cabin and kinda see how things are going. Okay? I don't want to hurry 'em, but I'd like to do it in another, oh, ten minutes." Mendenhall went into the cabin to check the situation there while McBroom and Beebe discussed the landing and evacuation. The air traffic controller put the plane into the next holding pattern. A moment later, Mendenhall returned to report that there were around 3,000 pounds of fuel left. The controller asked how much waiting time the crew needed. Copilot Beebe responded, "Yeah, we have indication our gear is abnormal. It'll be our intention in about five minutes to land on (runway) two eight left. We would like the (fire) equipment standing by. We've got our people prepared for an evacuation in the event that should become necessary."

When Portland air traffic control asked for details of the number of people on board and the volume of fuel upon *landing*, McBroom answered, "One seventy two and about four thousand, well, make it three thousand pounds of fuel and you can add one seventy two [passengers] plus six infants."[71] However, this figure represented the fuel remaining at

---

71 This information was incorrect in terms of not only the fuel volume but also the number of passengers. There were 181 passengers and eight crew members and, at the time the report was made, the fuel had already dropped to 3,000 pounds.

that time, not on landing. As it was, the figure of 4,000 was already an overestimate. There were at least another 10 minutes to go before landing. The figure quoted by McBroom should have been 2,200 pounds less, indicating a remaining reserve of not even 1,000 pounds.

At around 6:06 p.m., Wheeler entered the cockpit and said the cabin was now ready. "Okay we're going in now, we should be landing in about five minutes," replied McBroom.

At the same moment, Beebe said to McBroom, "I think you lost [engine] number four, Buddy," before turning to the flight engineer: "Better get some crossfeeds open there or something."

McBroom, however, had failed to notice the engine failure. He was still talking to Wheeler. Beebe turned to him again. "We're going to lose an engine, Buddy."

"Why?"

"We're losing an engine."

"Why?"

"Fuel."

Beebe turned back to Mendenhall. "Open the crossfeeds, man."

At nearly the same time, McBroom said to Mendenhall, "Open the crossfeeds there or something. [My fuel gauge is] showing a thousand or better." "I don't think it's in there," answered Beebe.

Pilots and flight engineers each have their own tank indicators, which can display slightly different values. When the fuel level is very low, technical issues can affect the accuracy of the tank indicators.[72] "Okay, it, it's a ... " started McBroom. Beebe said, "It's flamed out."

McBroom asked Portland air traffic control to issue immediate instructions for landing and was given permission for a visual approach to runway 28 left. Flight engineer Mendenhall reported that engine number three was also on the brink of failing. "It [the fuel gauge] is showing zero." McBroom answered, almost pleadingly, "You got a thousand pounds, you got to. ... " In response, Mendenhall said, "Five thousand were in there, Buddy, but we lost it." "Okay," replied McBroom, while he and Beebe tried to restart engine number three.

McBroom told Mendenhall to give the cabin the signal for an immediate landing and asked the air traffic controller how far it was to the airport. "I'd call it eighteen flying miles," the controller replied.

"Boy," said Mendenhall, "that fuel sure went to hell all of a sudden. I told you we had four [thousand pounds]." "There's a kind of an interstate

---

72 The reliable range of deviation is between +/-200 and +/-400 pounds.

highway type thing along that bank on the river in case we are short," said McBroom. "That's Troutdale [a small airport] over there ... "[73] "Let's take the shortest route to the airport," said Beebe.

At 6:13 p.m. Mendenhall stated, "We've lost two engines, guys. We just lost two engines, one and two." "You got all the pumps on and everything?" asked Beebe. "Yep" was Mendenhall's brief reply. "They're all going. We can't make Troutdale," said McBroom. "We can't make anything," replied Beebe. "Okay, declare a Mayday," ordered McBroom.

Beebe called Portland tower. "United 173 heavy, Mayday, the engines are flaming out, we're going down, we're not going to be able to make the airport." It was the last radio transmission from United 173 (Figure 1.17).

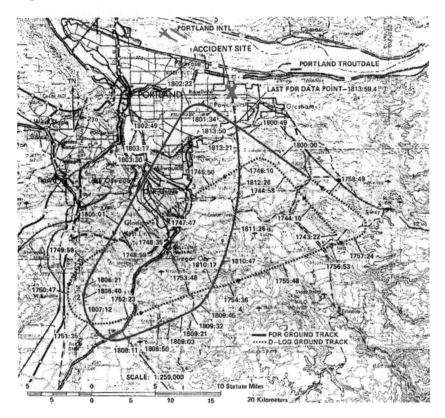

Figure 1.17 **Flight path of UAL 173**

---

73 A small airport that was not suitable for landing a large passenger jet and was only considered as an option in the event of an emergency landing.

McBroom and his crew managed to make an emergency landing. At around 100 feet above the ground, the DC-8 struck a wooded area six miles southeast of Portland Airport. Slowed by the trees, it came to a halt on the ground after 1,230 feet (Figure 1.18). Eight of the 181 passengers on board were killed by the impact. A further 21 were seriously injured. Mendenhall and Wheeler were also killed.

The NTSB published its accident report six months later. In essence, the report found[74] that the captain was responsible for the accident. He failed to heed information from his crew and check the fuel levels properly. The report also criticized Beebe and Mendenhall for failing to ensure McBroom was fully aware of the limited fuel supplies. Ironically, the investigation also revealed that the main landing gear was fully extended and ready for landing at the time of the impact. However, regardless of the responsibility borne by captain McBroom, the NTSB decided the true cause of the crash lay elsewhere. Two experienced crew members had failed to communicate a simple message to their superior, namely that the fuel was running low and the plane had to land.

Figure 1.18   **UAL 173 after the emergency landing**

---

74 NTSB (1979), p. 29.

We all, including the NTSB, see everything better in hindsight. For that reason, we should take a closer look at some of the key situations in this scenario. Among other things, it reveals one of the most treacherous causes of errors: misinterpretations based on assumptions and silent misunderstandings. In this case, we have McBroom, an experienced captain, who knows that if there are 5,000 pounds of fuel remaining, the plane will have to land within 20 minutes. He also knows that the flashing warning lights mean fuel is running low. In addition, we have Mendenhall and Beebe, two equally experienced crew members, who theoretically can assume their captain is fully aware of all aspects of the situation.

At some point, though, Beebe seems to become unsure as to everybody's awareness of their fuel situation. Yet, instead of voicing his concerns and pressing McBroom to land the plane, he asks Mendenhall about the amount of fuel remaining. Ten minutes later, he asks the same question. In both cases, he already knows the answer because, as copilot, he has his own fuel tank indicator. Yet again, he does not take the matter up with McBroom. Perhaps he hoped McBroom had overheard the conversation and interpreted everything correctly.

We are already familiar with this type of behavior from the case of JAL 8054. In that instance, flight engineer Yokokawa felt unable to address the mistakes made by captain Marsh and simply made general statements. Recall his phrase "Usually on yaw damper." This principle of hinting and hoping is a type of defense mechanism. It is a way of indirectly addressing errors made by a superior in the hope that he or she will get the message without taking umbrage.

However, McBroom's request for the landing weight and the given amount of fuel shows that he did not have a clear picture of the fuel situation at all. That is perhaps understandable in view of the imminent emergency landing. Instead of calculating the consumption for 15 minutes based on the existing 5,000 pounds – giving a reserve of around 1,700 pounds – he came up with a residual amount of 3,000–4,000 pounds, which he probably felt was a comfortable buffer. In an ideal world, when talking about their reserve, Mendenhall would have contradicted McBroom and stated the correct figure. Instead, he makes a series of vague comments. "Fifteen minutes is gonna really run us low on fuel here," he says. A little later, he notifies McBroom of the landing weight (i.e., the zero fuel weight of the plane plus fuel). However, the two men are basing their calculations on different fuel volumes – Mendenhall on the correct volume, McBroom on an incorrect one. Nevertheless, they assume they are in agreement. Shortly thereafter, when asked again about the fuel situation, Mendenhall gives the current figure of 4,000 pounds. Although Mendenhall's response is a little confused, McBroom could have understood the message. After all, Mendenhall was now talking about a figure that McBroom already had fixed in his head, namely something in the region of 4,000 pounds. After that, McBroom becomes distracted and starts discussing the emergency landing with his copilot. The next figure Mendenhall gives is 3,000 pounds. McBroom does not fully register this and seems to remain fixated on his figure of 4,000 or "make it three thousand pounds of fuel," mistakenly quoting this figure for the fuel reserve upon landing to the air traffic controller in Portland – even though it was the current amount and, for their landing reserve, far too high. The engine failures and the necessity of an immediate emergency landing – all within the space of a few minutes – must have come like a bolt from the blue.

So what went wrong? This time, fear of the captain cannot have played a role in what unfolded. Unlike in other cases we have examined, McBroom is not a man who insists on observing strict hierarchical distances between himself and those under his command. His tone is informal and his colleagues address him as "Buddy." He stays calm when a problem arises. There is very little age difference between McBroom, Mendenhall, and Beebe. Mendenhall calls Beebe and McBroom "boys." The atmosphere between the three men is relaxed. Nevertheless, shortly before the planned landing, both Beebe and Mendenhall must have noticed that McBroom no longer had an overview of the fuel situation. Even so, it did not occur to them to educate or correct him – and then it was too late to do anything and the fuel was gone. The question we inevitably have to ask is why did the pair – Mendenhall in particular – not speak up in time to ensure that McBroom was fully aware of the problem? After all, they were in the middle of an emergency situation that was getting more serious by the minute, plus the lives of everyone on board were hanging in the balance. Granted, there was not much to be done at the start of the problem. If two people are silently convinced of two different things – yet believe that they are in agreement and that what they think is right – then it will not, and actually cannot, occur to either of them to clarify, debate, or attempt to improve the situation. Such situations resolve themselves either by chance or are never actually identified. For that reason, it is more interesting to examine why the first information flow blockage occurred. After all, it triggered the subsequent series of vague, unhelpful communications.

This brings us to next situation in the cockpit of UAL 173. What we are talking about here is a commonplace internal attitude of lower-ranking individuals and the curious paradox whereby they believe that their bosses know better – but that they do, too. In everyday business environments, this type of attitude leads to an interminable cycle of statements such as "I could have told him that," "I knew that from the start," "The boss doesn't have a clue," and "I'm not surprised it hasn't worked." Yet, in the next moment, this can flip over into "He should know, he is the boss," "I'm not getting involved, that is above my pay grade," and "It has nothing to do with me." This behavior is not indicative merely of mental confusion. Rather, we are dealing with signs of silent antagonism between superiors and subordinates, signs that are sometimes papered over but can easily resurface. In this relationship, a superior – even someone as easygoing as McBroom – may on occasion consider his own knowledge to be privileged, his findings smarter, or he may not be greatly interested in what his next-in-command thinks. But maybe on the next occasion, the very next in command – even someone as friendly as Mendenhall – may let their superior take the fall because of this.

I am not suggesting that either Mendenhall or McBroom were intentionally behaving in this manner during the flight. However, when Mendenhall is instructed to calculate the landing weight in "about another fifteen minutes," he asks – whether to confirm the order or for no real reason at all – if McBroom means "fifteen minutes." McBroom does not respond to this. Perhaps he did not hear Mendenhall's question. Perhaps he felt no response was necessary or he was simply irritated by the question. All of us will be able to understand all three of these possibilities. This exchange is followed by McBroom's imperious order that Mendenhall should calculate 3,000–4,000 pounds on top of the zero fuel weight of the plane. In other words, just seconds after giving Mendenhall a task, McBroom effectively does it for him. With this in mind, it is possible that Mendenhall is left a bit annoyed. After all, not only has his question been ignored, but he also may feel McBroom has just treated him like a child by doing his job for him. It is also possible that – whether intentionally or semi-intentionally – he tries to punish McBroom later by not communicating the necessary information as clearly as required. "Not enough," he murmurs, almost to himself, regarding their fuel reserve. His comment "Fifteen minutes is gonna really run us low on fuel here," is slightly louder, but still falls short of making a concrete statement.

Of course, this possible sequence of behavior – disappointment/resentment/punishment – applies not only to the relationship between a first and second (or in this case a third) in command, but also to the way we generally deal with and react to other people. In the case of minor issues, we manage to swallow our annoyance. However, if we were in the shoes of Mendenhall, our subsequent reactions may be tinged with resentment. If we pursue this train of thought further, Mendenhall's sense of vindication – and even triumph – shortly before the emergency landing becomes more plausible. *That fuel sure went to hell all of a sudden.* This is accompanied by a reflection on that earlier moment that now provides him with satisfaction and a perfect alibi: *I told you we had four.* Mendenhall had tried to give McBroom the right information, but the captain failed to listen. Now, McBroom will have to answer for the consequences.

In business environments, the rule that "facts" should dominate our behavior is often cited, as if facts were sacred and should take preeminence over the emotions or feelings of the individual. If we follow this train of thought, Mendenhall should have forgotten his feelings and focused on the factual situation as being the only thing which counted. But he did not, independent of what he should have done. There are situations where the postulate of subordinating one's feeling to the rule of fact and reason is too simplistic a doctrine, particularly so, as it will not help in understanding human behavior. Undoubtedly, it is possible to suppress resentment, disappointment, anger, and irritation and swallow it down in the face of the "facts" (or at least appear to do so). Yet we can only guess what inner state of affairs these frustrated feelings may trigger; at least, we know fairly well what they trigger in ourselves and how long we choke on the feelings we have swallowed for the sake of so-called facts. In this respect, Mendenhall's behavior offers a nice example of the influence these emotions can have when we are under their spell and shows us how impotent reason or common sense may be at the times both forces compete.

The NTSB, too, began to think that new training requirements alone (such as constant monitoring of fuel indicators and clear distribution of tasks within the cockpit) would not be enough to prevent similar accidents in the future. After all, the pilots of UAL flight 173 had all received exactly these training courses and instructions on a regular basis. What was needed was a different, clearer mindset that even though it would put "the facts" at the forefront, it would do so in an emotionally secure environment, where all team members communicated as equals.

AVA flight 052: being too polite

Our first four cases featured copilots who found it difficult to make their views and opinions heard when dealing with their captain and superior.

This trend was also evident in a slightly less pronounced form in the cock-pit of UAL flight 173. Our next case involving AVA 052 introduces yet another perspective, where the copilot is explicitly required to actively participate in events. We will examine why things still did not work as they should have.

We begin with a closer look at the air traffic control system (Figure 1.19). Unless they are being flown according to visual flight rules, aircraft are under the guidance of air traffic control. This prevents mid-air collisions. Air traffic controllers use radar images to manage a clearly defined airspace. They direct the planes within this specific area to ensure that collisions are avoided. The instructions the controllers issue to the pilots over the radio must be followed unless they threaten the safety of the flight. For example, if a plane were instructed to fly through a storm cloud, the pilot is permitted to refuse to follow that order. In that case, the controller would try another route. If a pilot finds himself in a critical or emergency situation, he is obliged to inform air traffic control accordingly. The controller – sometimes an entire team of controllers – will then provide support up until the point of landing. Other planes may be diverted if necessary. Air traffic controllers often work in teams of at least two. One of them maintains radio contact with

Figure 1.19   **An air traffic controller's screen**

the planes in their assigned airspace. The other is responsible for coordinating operations with the air traffic controllers in charge of neighboring
airspace and managing the transfer of planes from one area to the next.
In the cases and accident reports we are examining, the controllers have
only one voice, even though they will change from station to station. It
reflects the experience of pilots, who only ever address one controller at
a time while in radio contact. Controllers are typically assigned to the
tower, approach control, departure control, and en route control. In the
case of AVA 052, we are dealing with en route controllers responsible
for the area over Norfolk, Virginia, and the controllers in charge of the
approach to New York's John F. Kennedy International Airport (JFK)
and JFK Tower.

Let us now look at our case. On January 25, 1990, a Boeing 707–321
belonging to Colombian airline Avianca (Figure 1.20) was being prepared
at Aeropuerto Internacional Eldorado in Bogotá, Colombia. Assigned
flight number Avianca 052 (AVA 052), the plane was due to fly to JFK[75] in
New York (Figure 1.21).

Figure 1.20 **Avianca Boeing 707–321**

---

75 All the following information is taken from the official investigation report by the
   NTSB; cf. NTSB (1991).

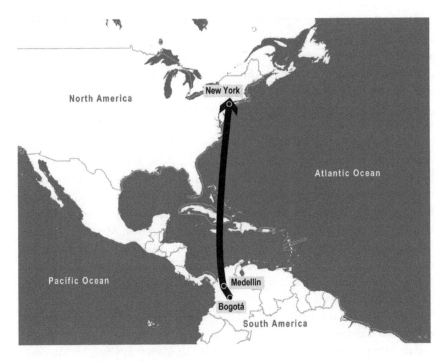

Figure 1.21 **Route of AVA 052**

The flight was scheduled to depart at 1:15 p.m. local time and land at JFK at 7:50 p.m. En route, it would make a stop at Aeropuerto Internacional José María Córdova, Medellín.

It was a routine flight for captain Laureno Caviedes-Hoyas (51) and flight engineer Matías Moyano (45). In contrast, copilot Mauricio Klotz (28) had only been flying longer international routes for a month.

During the preparations for the flight, Caviedes discovered that the autopilot in the Boeing 707 was defective. That meant he would have to fly the entire way manually, an extremely strenuous task. In principle, Caviedes could have refused to proceed with the flight due to this fault. However, he almost certainly would have run into trouble with the management of Avianca, since aviation authorities permit normal scheduled flights even without functioning autopilots.

Unfortunately, the defective autopilot was not the only problem. A large area of low pressure with rain and low, solid clouds covered the entire northeastern seaboard of the United States. The weather was particularly bad over New York City. The cloud base there was at 200 feet. Partial fog meant visibility was under a mile. On top of all this, there was a strong, very squally wind blowing from the south. The conditions were no better at Boston Logan International Airport, where the plane would divert if necessary. Although delays were likely, landing was nonetheless possible and permitted even in these conditions. In such poor weather, however, the intervals between planes landing are longer than normal.

At 1:10 p.m. local time, five minutes before the scheduled departure time, AVA 052 took off from Bogotá. After nearly an hour in the air, it landed in Medellín at 2:04 p.m. While more passengers boarded, the crew received an update on weather conditions in New York. Things had not improved. Captain Caviedes decided to fuel the Boeing 707 to its maximum possible takeoff weight. In addition to the calculated flight time of four hours and forty minutes, the plane now had enough fuel for a further two hours or so in the air. After around an hour in Medellín, captain Caviedes got the Boeing 707 off the ground at 3:08 p.m. There were 149 passengers and nine crew members on board AVA flight 052. Klotz took over responsibility for the radio communications. Unlike captain Caviedes, he spoke English well.

The flight ran completely smoothly to begin with. In hindsight we know that there was one slightly odd thing. During the remainder of the flight, and despite the poor weather, the crew failed to request any further information about the weather or traffic conditions at JFK airport, even though it would not have been difficult to get this information. At around 5:30 p.m., AVA 052 was abeam of Miami, where the local dispatch station could easily have provided the crew with the updates they needed. As it was, no one in the cockpit was aware that several planes had had to abort their approaches to JFK and try again, or that the delays were now longer than initially expected.

At 7:04 p.m., as the plane was over Norfolk, Virginia, the Air Route Traffic Control Center (ARTCC) informed the crew that their arrival in New York would be delayed. AVA 052 was sent into its first holding pattern. The waiting time was expected to be 19 minutes. No one was worried. Flight AVA 052 flew its next holding pattern over Atlantic City. This lasted 30 minutes. By the time the plane was able to resume its flight toward New York, it was already 8:09 p.m. Copilot Klotz then received radio instructions to move to an altitude of 19,000 feet over navigation point CAMRN,[76] around 38 miles south of JFK. The crew could expect the next set of instructions at 8:30 p.m.

By this time, captain Caviedes had been flying the plane manually for five hours. As a result, he probably lacked the concentration required to also focus on the traffic situation. Due to his rather poor English, Caviedes

---

76 CAMRN is a navigation waypoint on an air traffic route. Navigation waypoints are always identified by five letters.

got Klotz to repeat all the air traffic controllers' notifications in Spanish. At 8:17 p.m., AVA 052 was informed that it would remain in the holding pattern until 8:39 p.m. At 8:31 p.m., the crew was instructed to descend to 11,000 feet. Four minutes later, the controller asked whether visibility of 2,400 feet was acceptable for the approach for landing. Klotz confirmed it was. After another eight minutes, Klotz hesitantly asked air traffic control if the flight had permission to begin the approach.

New York ARTCC: "Avianca zero five two heavy, go ahead."

Klotz: "Thank you, sir, you have any estimates for us?"

ARTCC: "Avianca zero five two heavy, might be able to get you in right now, stand by one."

Klotz: "Thank you."

ARTCC: "Avianca zero five two, we just got off the line – it's indefinite hold at this time. Avianca zero five two, turn left heading zero nine zero. Join the Deer Park [radio beacon] two twenty, one radial, hold at CAMRN, maintain one one thousand."

This latest delay meant that the flight would land over an hour late. As the plane was now starting to run low on fuel, diverting to Boston was no longer an option.

ARTCC: "Avianca zero five two heavy, expect further clearance time zero two five."[77]

Klotz: "Zero two zero five, – well, I think we need priority, we're passing [unintelligible]."

The "priority" requested by Klotz is not a standardized term in radio communications. Consequently, the ARTCC controller would have interpreted Klotz's communication as a non-urgent request. Had Klotz used the standard terminology and reported "minimum fuel" or "fuel critical," the controller would have grasped the situation straightaway.

ARTCC: "Avianca zero five two heavy, roger, how long can you hold and what is your alternate [airport]?"

Klotz: "OK, stand by on that."

---

77 The times supplied by the air traffic controllers are UTC times (+5 hours compared to Eastern Standard Time).

Again, the air traffic controller asked how long AVA 052 could continue to wait.

> Klotz: "Yes sir, we'll be able to hold about five minutes – that's all we can do."

It was now 8:46 p.m. The five minutes quoted by Klotz referred to the amount of fuel remaining. His calculations were based on the regulation that, at the point of landing, a plane must have enough fuel left for at least 30 minutes (around 5,000 pounds in this case). This guideline is familiar to us from the case of UAL flight 173. Given this minimum fuel level and the fuel required for landing, AVA 052 would have to begin its approach within five minutes at the very latest. However, the controller understood Klotz's reply to mean that AVA 052 had five minutes in which to fly toward its alternate airport.

> ARTCC: "Avianca zero five two heavy roger, what is your alternate?"
>
> Klotz: "It was Boston, but it is – full of traffic, I think."
>
> ARTCC: "Avianca zero five two, say again your alternate airport."
>
> Klotz: "It was Boston, but we – we can't do it now, we don't, we run out of fuel now."
>
> ARTCC: "Avianca zero five two heavy – just stand by."
>
> Klotz: "Thank you."

There is no information to indicate whether the cockpit crew had discussed the fuel situation by this point. We have to assume that, while in the holding patterns, both captain and copilot thought they would receive permission to land with time to spare.

At this point, the ARTCC controller was handling seven other planes in addition to AVA 052, all of them heading for JFK and all of them with a cockpit crew impatient to land. He transferred AVA 052 to one of his colleagues at approach control and either failed to hear or did not fully grasp the crucial message *we run out of fuel now*. When passing the flight to his colleague, he said that AVA 052 was able to wait another five minutes in the holding pattern.

A little later, the approach controller in New York told Klotz to expect instructions to begin the approach to runway 22L. Things seemed to be taking a turn for the better. At 8:54 p.m., however, the controller instructed AVA 052 to enter a holding pattern. Curiously, the crew accepted this order. At 9:02 p.m., AVA 052 finally received the eagerly awaited permission to begin its final approach.

Caviedes: "Ave Maria!"

Klotz: "But now it is completed, isn't it?"

Caviedes laughed.

New York Approach: "Avianca zero five two heavy, call approach one one eight point four. Before you go, there's a wind shear reported on final at fifteen hundred feet. It's an increase in ten knots, then again at five hundred feet of ten knots reported by a Boeing seven two seven. Good night."

The crew was relieved. However, the approach controller had mentioned wind shear.[78] Such winds can require a crew to abort an approach and force them to begin the process again. That would consume yet more fuel. Moyano alerted the pilots to the problem. "When we have – with thousand pounds or less in any tank, it is necessary to do – then the go around procedure is stating that the power be applied slowly and to avoid rapid accelerations and to have a minimum nose up attitude."[79]

Caviedes: "To maintain what?"

Klotz: "Minimum nose up attitude that means, the less nose-up attitude that one can hold."

Moyano: "This thing is going okay."

Klotz: "They are accommodating us ahead of an ... "

Caviedes: "What?"

Klotz: "They accommodate us."

Moyano: "They know that we are in bad condition."

Caviedes: "No, they are descending us."

Klotz: "They are giving us priority."

---

78 Wind shear refers to vertical or horizontal differences in wind speed and direction. These winds normally occur in connection with storms, but can also be caused by cold or warm fronts. For planes, wind shear is particularly problematic during takeoff and landing. Planes flying through an area of wind shear experience rapid and dramatic changes in flight speed. During the takeoff and landing periods, planes are already flying very close to their minimum speed. Any loss of speed can quickly cause a critical loss of altitude in close proximity to the ground. In the event of wind shear, it is therefore often necessary to abort the approach. However, as the problems are almost always localized, it is usually possible to successfully complete another approach after a short time.

79 With this statement, Moyano is trying to alert the other crew members to the minimum climb angle required after an aborted approach in order to prevent the fuel pumps – and then the engines – from failing. As he was probably reading from the manual and translating the information into Spanish, his comments are somewhat incoherent.

At 9:11 p.m., AVA 052 received permission for an ILS approach to runway 22L. While Caviedes concentrated on following the ILS instructions in heavy turbulence, Klotz tried to offer support with a series of brief remarks. At 9:16 p.m., around eight nautical miles (nearly 15 km) from the runway, Caviedes gave the order to deploy the landing gear.

> Klotz: "I think is too early now."
>
> Caviedes: "It is the minimum in order to fly twenty-five."[80]
>
> Klotz: "If we lower the landing gear, we have to hold very high nose attitude."
>
> Moyano: "And it is not very–"

Both Klotz and Moyano tried to convince Caviedes that it would be better to wait a little longer before deploying the landing gear. This would reduce the drag, require less engine power, and thus reduce fuel consumption. Flying with the landing gear deployed would use up more fuel. In addition, it would increase the plane's pitch angle. This could interrupt the flow of fuel in the tanks.

Although Caviedes allowed himself to be persuaded by Klotz and Moyano, he was obviously becoming increasingly nervous.

> JFK Approach: "Avianca zero five two, can you increase airspeed one zero knots?"
>
> Klotz: "Say again the speed."
>
> JFK Approach: "Can you increase your airspeed one zero knots?"
>
> Caviedes: "Okay, one zero knots increasing."
>
> JFK Approach: "Increase, increase."
>
> Caviedes: "What?"
>
> Klotz: "Increasing."
>
> Caviedes: "What?"
>
> Moyano: "Ten knots more."
>
> Klotz: "Ten little knots more."
>
> Moyano: "Ten little knots more."

---

80 Here, Caviedes is referring to the guidelines for flying the plane. As far as he is concerned, these stipulate that the landing gear should be deployed when the flap position is 25 degrees.

Caviedes: "One hundred fifty – here we go. Tell me things louder, because I am not hearing it."

The approach flight was very turbulent and Caviedes needed to focus his strength on stabilizing the Boeing 707 in the strong, gusty wind. The controller's instruction to increase the plane's speed by ten knots was intended to increase the distance between AVA 052 and the planes behind it. For Caviedes, it was just one more thing to deal with. Shortly after, the controller from JFK Approach instructed the crew to switch to the JFK Tower frequency.

Caviedes: "I'm going to approach one hundred and forty. It is that what he wants or what is the value he wants?"

Klotz: "One hundred and fifty – we had one hundred and forty and he required ten little knots more."

Caviedes gave the order to deploy the landing gear and set the flaps to 40°. This was an unusual decision as it would further increase the drag, making it more difficult to increase the speed, to say nothing of requiring more engine power and consuming more fuel. It is possible that he was now too tired to concentrate properly.

Although the plane was then ready to land, it was still flying too slowly. The controller in the tower once again asked the crew to increase the speed to 150 knots, as the next plane in the line was drawing closer. Caviedes managed to get the speed up to 145 knots. AVA 052 was cleared to land at 9:19 p.m.

Just after that, Caviedes ordered the angle of flaps to be changed from 40° to 50°. Neither Klotz nor Moyano contradicted him.

At 9:22 p.m., Klotz advised that the plane was at 1,000 feet, slightly under the glide slope.

Klotz: "Below glide slope. Glide slope. This is the wind shear."

Moyano: "Glide slope."

At this moment, the ground proximity warning system (GPWS) sounded. A synthetic voice combined with an alarm warned the pilots to at least arrest their descent or abort the landing.

GPWS: "Whoop whoop – pull up."

Klotz: "Sink rate."

GPWS: "Whoop whoop – pull up."

Klotz: "Five hundred feet."

The warning signal sounded a total of 15 times. Caviedes continued his approach regardless. He wanted to land. Klotz and Moyano also failed to respond to the warning. Instead, Klotz called out the altitudes, just as if it were a normal landing. At 9:23 p.m., Caviedes finally spotted the approach lighting, but was still unable to see the runway. The GPWS continued to warn of a collision with the ground.

Caviedes: "Where is the runway? – The runway, where is it?"

GPWS: "Glide slope. Glide slope."

Klotz: "I don't see it. I don't see it!"

It was no use. Neither Caviedes nor Klotz were able to establish a visual on the runway. Caviedes aborted the approach.

Caviedes: "Give me landing gear up – landing gear up. Request another traffic pattern."

Klotz relayed this request to the tower controller. By this point, the fuel was almost gone.

JFK Tower: "Avianca zero five two heavy roger. Climb and maintain two thousand turn left heading one eight zero."

Caviedes: "We don't have fue–"

Klotz [repeating the controller's instructions for Caviedes]: "Maintain two thousand feet one eight zero on the heading."

Caviedes: "Tell him we are in emergency."

If a plane declares an emergency by using the word "emergency" or "Mayday," it is immediately given priority. Basically, the crew is then able to determine where they want to land and what approach they wish to use, provided this will not endanger anyone else. Air traffic control ensures the space required is cleared.

Klotz [to JFK Tower]: "We'll try once again – we're running out of fuel."

JFK Tower: "Okay."

Caviedes: "What did he say?"

Klotz: "Maintain two thousand feet, one-eighty on the heading. I already advised him that we are going to attempt again because we now – because we can't..."

Caviedes: "Advise him we have an emergency. Did you tell him?"

Klotz: "Sí señor. I already advised him."

JFK Tower: "Avianca zero five two heavy, continue the left turn, heading one five zero maintain two thousand. Contact approach on one one eight point four."

Klotz: "One one eight point four."

The next problem came once the crew switched frequencies to JFK Approach. Instead of instructing an approach, the controller there told AVA 052 to climb to 3,000 feet.

Caviedes: "Advise him, we don't have fuel."

Klotz: "Climb and maintain three thousand and we're running out of fuel, sir."

JFK Approach: "Okay, fly heading zero eight zero."

Caviedes: "Did you already advise him that we don't have fuel?"

Klotz: "Sí señor. I already advise him. We are going to maintain three thousand feet and he's going to get us back."

JFK Approach: "Avianca zero five two heavy, I'm gonna bring you about fifteen miles north east and then turn you back onto the approach. Is that fine with you and your fuel?"

Klotz: "I guess so, thank you very much."

Caviedes: "What did he say?"

Moyano: "The guy is angry."

This was the last moment where it might have been possible to change the fate of AVA flight 052 (Figure 1.22). The air traffic controller asked if the plane had enough fuel for this extended vector. However, by now, Klotz was completely confused, saying "thank you" when he should have objected. Moyano, on his part, suddenly claimed the controller was annoyed, for which there was no proof whatsoever. Statements later made by a (surviving) flight attendant confirm that Moyano realized how serious

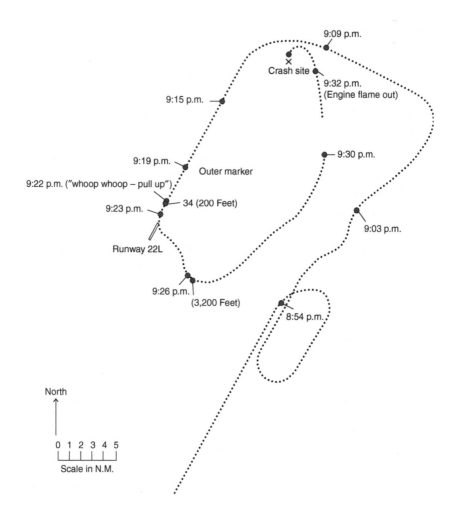

Figure 1.22 **Flight path of AVA 052**

the situation had become. During a brief visit to the cockpit, she saw him draw the edge of his hand across his throat and point to the indicators showing the fuel tanks were empty.

Despite the desperate situation, the crew prepared to attempt another approach.

Klotz: "We must follow the identified ILS."

Caviedes: "– to die. Take it easy. Take it easy."

Klotz: "When can you give us a final now, Avianca zero five two?"

The air traffic controller confirmed he had understood the question, but continued to let the plane continue its curve away from the airport. He did not reply to the question about the time, but was obviously aware that AVA 052 was experiencing problems even if he did not realize how serious the situation was. As a result, he gave the preceding TWA 801 flight permission to commence its approach before AVA 052.

Two minutes later, at 9:32 p.m., Moyano reported the first engine to fail due to a lack of fuel: "Flame out on engine number four."

Caviedes: "Flame out on it."

Moyano: "Flame out on engine number three. Essential [power] on number two – on number one."

Caviedes: "Show me the runway!"

The final act was about to unfold. Avianca 052 was still almost 20 miles away from JFK airport when the second engine failed. Caviedes looked in vain for the runway. Klotz radioed the approach controllers, but still failed to indicate an emergency.

Klotz: "Avianca zero five two, we just lost two engines and we need priority please."

JFK Approach: "Avianca zero five two, turn left heading two five zero intercept the localizer."[81]

Klotz: "Roger."

JFK Approach: "Avianca zero five two heavy, you're one five miles from outer marker, maintain two thousand until established on the localizer cleared for ILS two two left."

Klotz: "Roger Avianca."

It was the last radio communication from AVA 052.

In the cockpit, Caviedes desperately tried to bring the plane into the localizer of the ILS.

Caviedes: "That no – that. Did you select the ILS?"

Klotz: "It is ready on two."

This was the final sentence on the voice recorder.

---

81 Instruction to enter the localizer area of the instrument landing system.

JFK Approach: "Avianca zero five two, you have enough fuel to make it to the airport?"

JFK Approach [to JFK Tower]: "We're not talking to Avianca any longer. He's fifteen northeast of Kennedy."

JFK Tower [to JFK Approach]: "Umhm."

JFK Approach: "Okay, so if you get him he's nordo [no radio]. We don't know what his altitude – what his problem was. He last reported losing an engine."

JFK Tower: "Ah, wonderful."

JFK Approach: "Okay."

JFK Tower: "Thank you."

JFK Approach: "Avianca zero five two, New York. Avianca zero five two, radar contact lost."

JFK Approach: "We lost radar contact fifteen northeast of Kennedy with Avianca."

JFK Tower: "Thank you."

When engines one and two eventually failed, Caviedes went into a glide flight. It was too dark to find anywhere to make an emergency landing. Close to the small town of Cove Neck, Long Island, the plane clipped some treetops, struck a slope, and crashed to the ground (Figure 1.23). The cockpit and front section of the cabin were completely destroyed. The crash killed 73 passengers and crew, including the three members of the cockpit team. Eighty-four passengers and one flight attendant survived, but suffered serious injuries.

The official investigation report by the National Transportation Safety Board focused strongly on the communications between Klotz, the captain, and the New York air traffic controllers. The majority of the Board's criticism was directed at the ambiguous radio communications given by copilot Klotz.

In terms of the communications as a whole – both in the cockpit and between the various air traffic controllers – it is true that Klotz played an unfortunate role. However, what we are really dealing with here is a system accident with a chain – and chain reaction – of errors,[82] and rational decisions that were based on an incorrect context.

---

82 Cf. Helmreich, R.L. (1994); Perrow, C. (1999) also gives a comprehensive description of system accidents, see in particular pp. 124ff.

Figure 1.23   **Wreckage of AVA 052**

Although hindsight and reflections starting with "if only I had" and "if only he had" can change nothing about what actually happened, they are an unavoidable part of the error analysis process. They are not intended as reproaches or to apportion blame, but rather to show why these fatal mistakes happened and how they can be remedied and/or avoided. In this case, the aviation authorities needed to review the rule allowing crews to fly without a functioning autopilot, at least on longer routes. Flying manually for such a long time is so exhausting for the pilot that his reserves of strength and concentration may be completely drained. Another thing to be changed would be the handbook issued by Avianca. At the time of the investigation, it emerged that it was open to misinterpretation: in the event of fuel running low, it recommended requesting "priority," when the correct word would have been "emergency."

Nonetheless, the central character in this drama was undoubtedly Mauricio Klotz. He was too young, too nervous, and too inexperienced to cope with this crisis situation. Remember the age differences between the crew of AVA flight 052. Those of us over 28-years-old will certainly recall how old someone of 51 seemed to us at that age, and that at 28 we rarely gave instructions to a 51-year-old, particularly when he or she was also our superior. As to AVA 052 we have a young copilot who has been flying longer

international routes for a month and suddenly finds himself in an emergency
situation that only someone with the correct level of experience and support
would normally be able to deal with. We recall that Klotz is responsible for
radio communications with the various air traffic controllers from the point
the plane goes into the first holding pattern. After just one month on the job,
he would not yet have been familiar with the controllers' internal jargon and
staccato way of speaking. Nevertheless, he had to process the information
they provided, translate it, put it in the context of AVA 052, and respond to
it. He did all this more or less on his own. Quite apart from his exhaustion,
Caviedes' English was not good enough to be of help. As second officer,
Moyano did not get involved. In other words, Klotz was in over his head. As
if this was not enough, there was the fatal breakdown in communications.
Based on the information in his handbook, Klotz thought he had to ask for
"priority" in order to receive permission to land immediately. Consequently,
he believed the air traffic controllers were aware that his plane was in an
emergency and that all he had to do was wait for instructions to land.

The air traffic controller, on his part, might have thought that AVA flight
052 was trying to push its way up the line – which led to the next misunder-
standing, namely Klotz saying AVA 052 could hold on another five minutes.
He, of course, meant that the flight needed to land in five minutes. The
controller, on the other hand, understood these five minutes as a time buffer
for AVA 052 to fly to its alternate airport. After that, Klotz and the controller based their actions on their individual understandings of the situation.
We saw something similar in the case of UAL flight 173, where captain
McBroom and flight engineer Mendenhall as well as copilot Beebe followed
parallel modes of thinking. A more experienced, less overwhelmed pilot than
Klotz might have made sure that the controllers were aware of the emergency
situation. With everything that is being thrown at him, Klotz was barely able
to keep his head above water. As a result, the communication gap between
him and the tower became ever larger: the alternating air traffic controllers
continued to believe that AVA 052 was just another plane waiting to land,
maybe with a crew impatient to get on the ground, even if they needed to be
told to speed it up. In the meantime, Klotz believed or hoped that AVA 052
would soon be given permission to land. It is almost heartbreaking to read
how unsure he is transforming the 10 requested knots into inoffensive "little
knots" – anything to dilute the severity of the situation. Due to the voice
recorder, we know that at this point he had also started speaking too softly.
Klotz, in other words, had become disoriented and panicked.

When the ground proximity warning system sounded, it seemed as
though everyone lost their heads. Caviedes wanted to get on the ground at

last and started feverishly to search for the runway. Moyano seemed paralyzed, while Klotz was too confused to grasp the crucial final lifeline when JFK Approach asked, "Is that fine with you and your fuel?" He simply replied, "I guess so, thank you very much."

So what can the case of Mauricio Klotz teach us in terms of error analysis? It brings us to the organizational aspect of *empowerment*. According to current thinking, it is impossible to motivate promising employees without strong elements of empowerment. In the cases we looked at before AVA 052, we saw that the employees – that is, the second and third members of the cockpit crew – seemed to be motivated, even though they were not in the least empowered. They neither knew how to have their captains listen to them nor how to assert themselves. Instead, our copilots and flight engineers just froze, or sulked. Regarding AVA 052, the opposite is true. Even though Klotz is right at the heart of the action and encouraged to take control, he is unable to rise to this challenge. Though he is empowered, he does not know how to use his new role and can only fail in face of the overwhelming demands. Klotz, of course, was young and inexperienced. Still, he demonstrates that empowerment is pointless if an individual has not been equipped with the necessary knowledge and strengths. Sure, like any other pilot, Klotz had simulator training for emergency situations, every six months. Yet these sessions focused on flying skills, not on assuming a leadership role. Even for someone in a company, empowerment does not automatically render them *competent* and self-assured. That is particularly true if, like Klotz, they still lack the confidence that routine brings.

This is why, in addition to technical skills, the aviation industry finally began to train its copilots, flight engineers, and flight attendants to demonstrate a greater degree of courage in relation to their captains. They were taught to view themselves as the captain's colleagues, not his inferiors, and to assume the responsibilities that come with this level of empowerment. They did not learn their new role from one day to another. It took years until this new training, called CRM, became a success.

Part II Crew Resource Management

We all remember mistakes we would rather forget. Often they have caused us embarrassment so intense that we wished to disappear completely – preferably to a place where no one knew us. As I said in the beginning, mistakes are nothing we welcome, and embarrassment is one of the many unpleasant emotions we experience after having made them. Still, if they were not relegated to the most hidden corners of our brains, we might learn a lot from them. The lesson may not always delight us, but it will at least give us the chance to comprehend why we – stupidly, mistakenly, or maybe only subconsciously – did what we did. No matter how big the blunder and no matter how much we wish someone else was responsible for it, we and others can still learn from the mistake – particularly if it was a major one.

With regard to the size of mistakes, Charles Perrow offers one of his many impressive examples of errors and the accidents they have caused.[83] This one refers to the construction of the Grand Teton Dam, which began in 1972. It was to be built on a tributary of the Snake River in eastern Idaho. "In December of that year a group of geologists from the US Geological Survey were working in the area, and became concerned about the dam since they had evidence that the area was seismically active – that is, had recently had earthquakes. One of the geologists drafted a memorandum intended to alert their superiors in the Geological Survey and officials in the Bureau of Reclamation of the danger. [...] The memo finally reached the Bureau of Reclamation six months after it was first drafted. [...] Though there is no evidence in the house committee report relating to this, I think the Geological Survey's action (or lack of it) is readily understood. [...] The Bureau of Reclamation had by then spent $4,575,000 on the designs and the initial construction of a dam in a particular spot. An agency would not normally run to an adjoining and cooperating agency, and say, 'A mistake has been made. Four and a

---

83 Cf. Perrow, C. (1999), p. 23ff.

half million dollars have been wasted because of it. Find another spot, or redo your work even though it may double your costs.'"[84]

We can probably identify with that kind of reluctance, but for the Bureau of Reclamation, it would have been better if they had mentioned it. As no one did, the building continued. "But in 1973, with the dam half built [...] the Bureau found that on the right side of the damn the cracks were caves, large enough for a person to walk through. [...] Seepage could be a serious problem, the memo [this time of the Bureau] noted. But they went ahead, filling the reservoir at the standard rate of one foot a day as set in the original design."[85]

"On June, 3 1976, the leaks were located downstream of the dam, and a third was found the next day. The project engineer, Mr. Robinson, said he was not worried; the leaks were running clear, and clear leaks are common in earth dams."[86] More leaks appeared. The last one "leaked 22,000 gallons a minute. An hour and a half later, the final leak appeared, in the same area. As it grew, it sucked material from the embankment through an ever-widening hole. Earth-moving crews tried to fill in the hole, but a whirlpool grew even larger, and just after they abandoned their equipment and fled, it sucked the equipment in. Warnings were flashed to the people below the dam. Mr. Robinson, one assumes, was now worried. At 11:57 a.m. in June 5, 1976, the dam was breached. [...] More than 100,000 acres of farmland were destroyed and 16,000 head of livestock were lost. Total property damage was estimated at the time to be over $1 billion."[87]

How large was the initial sum the Bureau of Reclamation might have wasted? Perhaps we whistled while reading it, but compared to the $1 billion spent in the end, it was trivial. Admittedly, to go and confess early on that about $4.5 million had been lost required a brave heart. But after the initial shock, denial, uproar, and resignation, people might have cooled down, started to calculate the worst-case scenario and its costs, and could have seen that the latter would certainly hurt more than $4.5 million. Even better: what if the Bureau of Reclamation had had a system to deal with errors – even one of such dimensions? This would have avoided the eventual blame game and scapegoating.

A similar situation arose during the construction of two steelworks belonging to the ThyssenKrupp Group. At the height of the steel boom

---

84 Ibid., p. 235.
85 Ibid., p. 236.
86 Ibid., p. 237.
87 Ibid., p. 238.

in 2005, the plan was to build two new steelworks – one in Brazil and one in the United States – at a cost of around €2 billion. Then the price of steel dropped. At the same time, emerging nations increasingly started developing their own production capacities. Although this was a blow to ThyssenKrupp, it was still within the scope of those uncertainty factors that must be calculated into all planning processes. Worse was the fact that the site in Brazil had been unsuitable from the outset. Apart from the lack of a proper infrastructure, the site chosen was a marsh, which was proving to be a major problem. Whether they realized it immediately or later on, the experts from ThyssenKrupp learned about it at some point. Despite hefty additional costs, due to the insufficient supervision of subcontractors, both plants were completed. Perhaps no one was willing to take responsibility for the write-offs if the work had been canceled. By the time both plants were finished, construction costs had spiraled from an original budget of €2 billion to €12 billion.[88]

Most of us probably also remember the Barings Bank case, a prime example of the failed trading strategies of a single trader that led to the bank's collapse in 1995.[89] Mistakes were made, but no one seemed to learn from the experience and take steps to monitor such business dealings in the future. In May 2012, it emerged that US bank JPMorgan Chase & Company had taken serious risks in trading credit default swaps over a considerable period of time. This ultimately resulted in losses of more than $5 billion. Things were even worse at UBS and Société Générale. In these cases, however, the problems were due to fraudulent practices by rogue traders. What all the cases have in common, though, is that initially all of the traders made large profits so that the early warnings fell on deaf ears. For most banks, all that mattered was their notion of success, but the losses ultimately threatened their existence.

The construction of the new Berlin Brandenburg Airport (BER) provides another example. Shortly after German reunification, BER was planned as an impressive and prestigious project consolidating Berlin's three airports – Tegel, Tempelhof, and Schoenefeld. Although the contract was originally awarded to the experienced engineering firm Hochtief AirPort, the federal states of Berlin and Brandenburg subsequently took over construction responsibilities to save money. The project suffered a number of delays that exceeded the norm. The airport was due to be officially opened on June 3, 2012, and flight plans had been finalized far in advance. To keep

---

88 Cf. Sturbeck, W. (2012).
89 Bernard, A. et al. (2002), Soane, E. et al. (1998), Stonham, P. (1995).

this deadline, the construction management team had introduced a "traffic light" system to monitor progress. Green indicated that work was going as planned, yellow showed delays, and red warned of more serious problems. As the delays grew longer, the subproject managers switched their signals to green or sometimes yellow, even if this had nothing to do with the real state of affairs. Just one week before the planned opening and related festivities, building authorities refused to grant BER an operating license because fire safety systems were not functional. After a week of checks, it became clear that the situation could not be rectified. BER would be unable to open before fall/winter 2015.

Needless to say, the costs had gone through the roof. By May 2013, the original sum of €630 million for the new terminal building had sky-rocketed to €1.2 billion. This figure does not even include anticipated legal action for damages from airlines and retailers, who had counted on starting operations at BER in June 2012.[90] Certainly, the project manage-ment team is not the only one to blame for the additional costs. Yet, had the problems been identified earlier, hundreds of millions in cost overruns could likely have been saved from a total bill of more than €2 billion (as of May 2013). The dawning insight that the new airport building would not be big enough in the end was never addressed either, as any such report would have pointed to the costs necessary to make the extensions. As it is, the new BER airport will perform at the same level as the older "too small" Tegel Airport, which it was supposed to replace. In the meantime, thoughts have turned to erecting an enormous tent as an additional tem-porary terminal for BER.

Does that mean that if we only learned from our mistakes that they would have a silver lining? That is not always the case. There was no upside for those who died or were injured in the accidents I described in the first part of this book. If you have been physically impaired because someone else was exhausted, impatient, inexperienced, or consulted a faulty manual, it is of little comfort to know that this will be remedied in the future. The German taxpayers, too, will not find consolation in knowing that people might learn from the BER disaster. If there is an upside, it is this: hopefully, those involved will analyze what happened and aim not to repeat the mistakes. Anything else would lead nowhere and help no one.

Considering this, it is surprising how few tools, if any, we have available to check if we are making mistakes. We can try to self-correct. The problem

---

90 Cf. Küpper, M. (2012).

is that we often do things without knowing that we do them, independent of whether they are right or wrong. Our captains and copilots did not consciously observe themselves *not* listening to their colleagues or making the wrong decisions. Often we have to rely on those closest to us to kindly, angrily, or gleefully point out our errors. Even when they do it in an annoying way, we should be grateful, if only because it prevents us from becoming complacent. By comparison, think of people who live or work alone and discuss matters with themselves and how agreeable a discussion partner the "self" usually is, and how kindly it tells us that we are right, as always. But our self is a deceptive friend. In the end, it deludes us into believing that we are infallible, which may have been the cause for the inflexibility we witnessed in some of our captains. As for those not necessarily close to us – I am talking about coworkers and employees – we have seen that they often prefer to remain silent rather than tell us we are wrong.[91] Similar to the copilots in part I when dealing with their captains, they may have learned to fear our negative responses.[92]

All the cases we have examined so far have one thing in common: certain traits, mistakes, moods, and physical states have resulted in fatal errors. These errors might have been avoided had those involved openly communicated with one another as equals. Gradually, the aviation industry started to realize this, too. The first report issued by the US Air Force Inspector General on the lack of communication within cockpit teams was published as early as 1951 under the title "Poor Teamwork as a Cause of Aircraft Accidents."[93] The report was based on 7,518 accidents that occurred between 1948 and 1951.[94] Even back then, the authors suggested teamwork training programs as means to address the problem. However, these were never implemented. It took 27 years before this proposal was revisited, ultimately resulting in the first CRM concept.

Prior to me introducing CRM, let us briefly look at the distribution of tasks in the cockpit. From the very early days of aviation until the end of World War I, virtually all aircraft were flown by solo pilots. It was only at the start of the 1920s and the advent of increasingly large planes – like the W8 from Handley Page (Figure 2.1) or the Vickers Vimy – that more members joined the crew.

---

91 This phenomenon has been the focus of several empirical studies, for example Rosen, S. and Tesser, A. (1970); Conlee, M.C. and Tesser, A. (1973); Roberts, K.H. and O'Reilly, C.A. (1974); Noelle-Neumann, E. (1974); Milliken, F.J. et al. (2003); Zao, B. and Olivera, F. (2006).

92 Edmondson, A.C. (1996); Redding, W.C. (1985).

93 Kern, T. (2001), p. 6.

94 Ibid.

Figure 2.1   **The W8 from Handley Page took to the skies in the 1920s**

In the beginning these new members simply played a supporting role: flight mechanics focused on the engine operation, navigators took charge of determining flight path and location, while radio operators were responsible for communicating with teams on the ground. All were there to serve the captain and took their instructions from him. Copilots did not become a fixture in the cockpit till larger planes like the DC-3 (Figure 2.2) or Junkers JU 52 came onto the scene in the early 1930s. Quite apart from the fact that planes were now too complex to be flown solo, the second pilot also provided reassurance for passengers, who no longer had to fear being left to their fates if the captain happened to become indisposed during the flight.

Captains who had earned their stripes in the pioneering days of aviation did not take kindly to having a copilot. They were used to making their own decisions and, at best, tolerating the presence of the support staff mentioned above. For them, copilots were a troublesome burden. The best they could wish for was that they would watch and assist, but stay out of the decision-making process. The copilot was "the idiot on the right who'd get everybody into trouble if not constantly watched."[95] Just as there was a definite pecking order in terms of status, the salaries of the various members of the cockpit crew also differed, with the captain earning much

95 Morgan, L. (1983).

Figure 2.2   **DC-3 in the mid-1930s**

more than the others. The rest of the team accepted this situation. Most of them were just happy to have finally made it into the cockpit. Now they had an opportunity to gain the flying experience they needed for their future careers in aviation. More specifically, the copilots hoped to one day become captains themselves, while flight engineers, navigators, and radio operators had to build up their flying experience before they could switch to bigger planes – say, from a DC-3 to a DC-4. Although the crew had to work very closely together, little attention was paid to the *cooperation* between them. The training given to pilots and the way they worked was very much determined by the image that dominated the very earliest days of flight. The leadership model revolved around the captain being the central figure in the cockpit. A copilot who had moved up the ranks to captain due to his experience and length of service acted no differently. He simply enjoyed the power he had dreamed of for so long. In cases where the supremacy of a captain turned into downright tyranny, the crew had no choice but to bite their tongues and just hope they would be assigned a different captain next time around.

Despite their elevated status, captains still had to go through a challenging series of trainings and tests. In the 1940s, the US Civil Aviation
Authority (the forerunner of today's Federal Aviation Administration) stipulated precise intervals at which captains of commercial aircraft had to be
tested on their knowledge and ability, particularly in emergency situations.
Regardless of how egomaniacal their behavior might be on occasion, they
were undoubtedly competent, experienced pilots.

There was another reason why the issue of cooperation in the cockpit
remained unresolved for such a long time. Until the mid-1960s, the technical reliability of aircraft continued to be the main bugbear of the aviation
industry. The piston engines still prevalent at that time were notoriously
unreliable, and accidents were frequently caused by engine failure. Radio
navigation systems were also not yet properly developed. This resulted in
some absurd crashes, including two separate incidents with KLM planes
in Cairo, for example: a DC-6 crashed in 1958, killing one person, namely
the copilot; in 1961, an accident involving a Lockheed L-188C Electra II
killed 20. In both cases, faults in the radio navigation system caused the
planes to deviate from their stipulated paths during the approach to landing, leading them to smash into sand dunes at low altitude.[96] Consequently,
it seemed logical to focus almost exclusively on technical aspects when
conducting the accident analyses. The international aviation authorities,
aircraft manufacturers, and airlines applied the same logic when planes
with jet and turboprop engines replaced piston engine models in the mid-
1960s. They expected this shift to automatically reduce the number of
accidents – and were very much mistaken; we have seen that from the
cases we have looked at so far.[97] That this expectation remained unfulfilled
has been confirmed not least by the results of black box analyses. Flight
data recorders and voice recorders – known as "black boxes," even though
they are actually bright orange – were fitted into larger passenger planes
starting in the mid-1960s. Intended to aid in the understanding of air
accidents and improve flight safety, they continuously recorded data on
speed, altitude, and engine performance, as well as cockpit conversations.
The accident analyses the boxes eventually helped to produce revealed that
technical issues played a secondary role in air accidents. It emerged that
more than 70 percent of all accidents were caused by "human error" and/
or pilot error[98] (Figure 2.3). These findings were found surprising, seeing

---

96 Bartelski, J. (2001), p. 77ff.
97 Weener, E. (1992).
98 Helmreich, R.L. and Foushee, H.C. (1993), pp. 5–6.

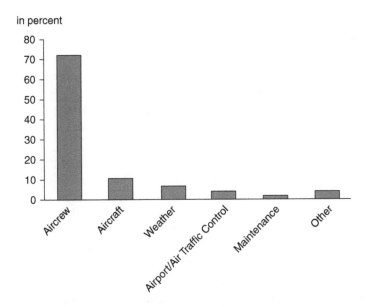

Figure 2.3 **Causes of air accidents between 1959 and 1989**

as the pilots involved in the accidents were all experienced professionals. To put it bluntly, highly skilled crews had managed to cause fully operational planes to crash. Finally, the authorities turned their attention to leadership behavior in the cockpit. As the National Transportation Safety Board (NTSB) realized after the United Airlines 173 crash, that was the real crux of the matter.

Until the end of the 1970s, the flight crew training required by the FAA in the United States, and corresponding authorities elsewhere, was specifically tailored to the individual areas of expertise (i.e., pilots, flight engineers, radio operators, and navigators).[99] However, following a systematic analysis of air crashes, NTSB came to the conclusion in 1979 that the number of accidents was not decreasing, despite all the training being given to the crews. It therefore became clear that continually improving the skills of flight crews was not the answer. In fact, some airlines were already ahead of their time. Pan American, for example, began introducing training designed to improve cooperation within its cockpit crews as early as the mid-1970s.[100] However, the conceptual framework needed to support these measures was still missing.

---

99 Hackman, J.R. and Helmreich, R.L. (1987), p. 289ff.
100 Helmreich, R.L. and Foushee, H.C. (1993), p. 7.

A new approach

The NTSB's analysis of the UAL 173 case shed even more light on the problem in the cockpit. As far as the authorities were concerned, neither flight engineer Mendenhall nor copilot Beebe had felt able – for whatever reasons – to explain to the captain that he had miscalculated the remaining fuel reserves. Consequently, the captain did not – could not – grasp just how precarious the situation had become.

After the evaluation of the accident data, but before the publication of the final report, a workshop was organized by NASA in conjunction with the FAA, the US Air Force, the aviation industry, and major international airlines including United Airlines, American Airlines, British Airways, JAL, SAS, and Swissair. The title of the workshop was "Resource Management on the Flight Deck."[101] It took place in June 1979, immediately after the publication of the final report on UAL 173. Using existing studies on the subject,[102] work began on carrying out a systematic analysis of cooperation in the cockpit. This put the dynamics within the cockpit team under the spotlight. In addition to numerous aviation experts, the working group included psychologists such as Robert Helmreich and J. Richard Hackman. The experts not only analyzed accident reports but also actively examined cooperation in the cockpit. Cockpit simulators were used to create realistic conditions in which crews could be confronted with all sorts of emergency situations. This, in turn, allowed the experts to gain insights into cooperation between the crew members.

In 1976, Hugh Patrick Ruffell Smith, from the Ames Research Center of NASA,[103] developed a simulator scenario for a Boeing 747 flight from Washington DC to London via New York. A total of 20 crews – each with captain, copilot, and flight engineer – had to "fly" this simulated exercise. The crews – all of them experienced pilots who flew normal scheduled flights – were only told that they must complete a transatlantic flight and that they should expect problems to arise. Ruffell Smith had designed the simulation in such a way that the first section from Dulles International Airport in Washington to JFK in New York would illuminate cooperation between the various crew members under normal flight conditions. In addition, the crew would have an opportunity to get to know one another.

---

101 Cooper, G. et al. (1980).
102 See, for example, Edwards, E. (1972, 1975), Ruffell Smith, H.P. (1979) and Hackman, J.R. and Morris, C.G. (1975).
103 Ruffell Smith, H.P. (1979).

However, autopilot number one was programmed to fail during this initial section of the flight.

The real challenge awaited the crew in the second phase of the flight. The cooperation and consultation between them was put to the test as early as the takeoff stage at JFK, with the plane at almost its maximum takeoff weight, marginal wind conditions,[104] and departure instructions that were changed at the last minute.[105] Following a routine ascent abeam of Boston, the simulation triggered a blockage in the oil filter in engine number two, leading to a drop in oil pressure. However, this problem was initially only visible on the flight engineer's indicators. Warning lights were only activated on the captain and copilot's instrument panel once the pressure reached a critical level. At this point, all three crew members had to realize they needed to shut down engine number two, as instructed in the quick reference handbook. If the oil pressure was too low, then parts of the engine would not be sufficiently lubricated. In the worst-case scenario, this could cause the engine to explode. Once they had shut-down engine two as required, the crew knew that they would be unable to continue the flight to London under these conditions. They either had to turn back or set course for an alternate airport. Because it was impossible to maintain the cruising altitude with the three remaining engines, they also had to agree upon a lower altitude with the air traffic controller. Furthermore, they needed to dump part of their fuel supply, as the plane was still carrying too much weight to land at that point. The weather conditions in the simulator were such that the only possible option was to return to JFK in New York and land there.

As the crew was dumping fuel into the air and preparing for landing, hydraulic system number three[106] failed, causing autopilots two and three to fail, too. After that, the Boeing 747 could only be flown manually, with the asymmetric thrust caused by the shut-down engine creating even more stress for those in the cockpit. Meanwhile, the weather conditions had changed and captain and copilot now had to contend with a strong cross wind during landing. They had to deal with all of this while constantly being interrupted by instructions from air traffic control and notifications from the cabin crew. The latter informed the captain that the passengers were annoyed at being told that the flight was returning to JFK. Although this concentrated series

---

104 A tail wind or strong cross wind combined with a short runway, for example.

105 The crew normally receives these before the engines are switched on, so that they have enough time to prepare. In this instance, this would intensify the crew's work prior to takeoff.

106 Large commercial aircraft have several hydraulic systems to ensure adequate redundancy should one system fail.

of incidents may seem exaggerated at first, the crew members participating in the simulation said the scenarios were realistic, albeit challenging.

Interestingly, the subsequent evaluation of each crew's performance revealed very different results. Some crews had worked together effectively and landed the plane safely, while others made mistakes that, in normal circumstances, would have put the flight at risk. Each time, the errors could be traced back to poor interaction. In none of the cases were they down to a lack of technical knowledge or insufficient flying ability.

Another study[107] about communication in the cockpit showed that the interaction between the crew members had an influence on performance. Crews that communicated less were more likely to make mistakes than those who spoke to each other frequently. At the same time, it also emerged that teams coped well with tricky situations and made fewer errors if they *constantly* updated one another on the flight status. For regular flights, 20 notifications per minute are deemed normal. In emergency situations, that figure rises to 35 per minute for crews who work well together. In the case of United Airlines 232, as described in part III, the crew reached 60 notifications per minute at times.[108]

A large-scale NTSB accident analysis covering the period 1978–1990 examined the role that captains played in errors made by crew members. It revealed that, in more than 80 percent of accidents, the captain was the pilot at the controls.[109] Theoretically, we might have expected the highly experienced captains to make the fewest errors of everyone involved. However, after what we learned in part I, we know better and are aware of the reason for this phenomenon: the other crew members failed to alert the captains to any mistakes they (the captains) made. In contrast, the captains were quick to correct their copilots and flight engineers.

Ultimately, the analyses compiled in the NTSB and NASA workshop in 1979 confirmed the assumption that cockpit teams needed to learn to work together better in order to improve safety in the air. Although the plan was not to abandon the hierarchical structure with its assigned responsibilities, the role of the captain would no longer be sacrosanct. That might sound a bit half-hearted, and begs the question: who, then, is responsible for determining where specific roles begin and end? I will return to that point later. Nevertheless, the conclusions drawn at the time by the workshop participants were used to develop a new concept for cooperation in the cockpit. Initially called Cockpit Resource Management, the name was changed to Crew

---

107 Foushee, H.C. and Manos, K.L. (1981); later also Predmore, S. (1991).
108 Helmreich, R.L. (1994), p. 275.
109 NTSB (1994), pp. 38–39.

Resource Management (CRM) in the mid-1980s. The concept faced massive resistance for a time, particularly from older captains, but also copilots, and flight engineers. However, after 10 years, it had become established.

The development of CRM

The realization that more attention had to be paid to cooperation in the cockpit in order to improve safety was not a new one. Following on from the "Poor Teamwork as a Cause of Aircraft Accidents" report mentioned previously, aviation psychologist Elwyn Edwards investigated the significance of the human factor in the operation of aircraft at the beginning of the 1970s and summarized the results in a SHEL model (see Figure 2.4).[110]

S = Software (procedures, operations manual, etc.); **H** = Hardware (cockpit layout, aircraft design, etc.); **E** = Environment (weather, day/night, unfamiliar aerodrome, busy airspace, etc.); **L** = Liveware (the person or people).

Figure 2.4 **SHEL model according to Edwards**

---

110 Edwards, E. (1972).

SHEL stands for the four core elements of a flight: software (programs, documents, and guidelines); hardware (the plane and all of its components, including instruments, radio equipment, even the emergency axe); environment (environmental conditions such as weather or other aircraft); and liveware (the individual crew members). Edwards focused predominantly on the interactions between the liveware, or flight crew. NASA referred to his findings when drawing up its CRM.

Among other things, Edwards examined the authority relationships between the captain and other crew members.[111] He concluded that it was the responsibility of the captain to ensure that this Trans-Cockpit Authority Gradient (TAG), as he called it, was neither too high nor too low (Figure 2.5). However, Edwards' work did not immediately result in the introduction of a training program. That only happened once his findings were taken up by the NTSB and NASA.

Although the term "Cockpit Resource Management" was coined during the initial "Resource Management on the Flight Deck" workshop mentioned earlier, all that lay behind it was the idea of making cockpit crews more team-oriented. To bring this idea to life, those in charge turned to management literature for help. Even at that time, there were already comprehensive research findings on teamwork and team behavior, all covered by the umbrella term "organizational behavior." These findings included the work of Meredith Belbin, a psychologist who laid the foundations in this particular area and whose research came to play an important role in the aviation industry.

As mentioned earlier, those present at the NTSB/NASA workshop included Robert Helmreich, professor of psychology at the University of Texas in Austin, and psychologist Richard Hackman, then a professor

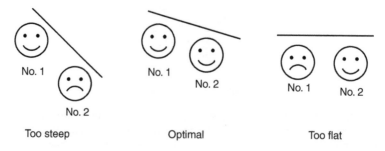

Figure 2.5   **Trans-Cockpit Authority Gradient – TAG according to Edwards**

---

111 Edwards, E. (1975).

at Yale University and later at Harvard University. Helmreich played a leading role in this process, both at this stage and in the ongoing development of the CRM concept in the years that followed. From the early 1980s, both Helmreich and Hackman worked on concepts to improve cooperation in the cockpit.[112] To gain a better understanding of specific situations, Hackman even took a pilot training course, flew regularly in the cockpits of commercial planes, and observed numerous crews.

Aviation authorities in the United States, the United Kingdom, and Germany, and the airlines United Airlines, KLM, British Airways, and Lufthansa all embraced and supported the development of the CRM concept. The crews themselves were less thrilled.

The first CRM generation: individual focus

Following the crash of UAL flight 173 in Portland, United Airlines was, in 1981, the first major airline to attempt to improve interaction among cockpit crew members, independently of the new studies and research.[113] Its program was based on the Managerial Grid concept developed by Robert Blake and Jane Mouton.[114] This management model depicts all the various permutations of staff and responsibilities. The aim was to use it to clarify the different forms of cooperation in the individual flight phases, hold accompanying seminars to identify the participants' leadership styles, and evaluate them in terms of their performance in the cockpit. The seminars were given by management trainers. Unfortunately, the captains, pilots, and flight engineers did not take the courses seriously, disparagingly dubbing them "charm school." To them, leadership issues and behavioral analyses seemed utterly trivial and inconsequential compared to the job of flying.[115] Others found it hard to accept a course that appeared to be a form of psychotherapy, while the captains simply saw it as undermining their authority.[116] In short, they rejected the CRM courses.

However, United Airlines – which played a leading role in this entire process – was one of those that did not stop at psychology-driven seminars.

---

112 See, for example, Hackman, J.R. and Helmreich, R.L. (1987).
113 For a comprehensive overview of the phases in the introduction of Crew Resource Management, see Helmreich, R.L. et al. (1999). For a European perspective that nonetheless very much reflects that of Helmreich, see Pariès, J. and Amalberti, R. (1995).
114 Blake, R.R. and Mouton, J.S. (1964).
115 Helmreich, R.L. et al. (1999), p. 2.
116 See Helmreich, R.L. et al. (1999), p. 21. The problems of CRM implementation are described in Byrnes, R.E. and Black, R. (1993) pp. 422–424.

The airline also expanded the range of situations covered by the half-yearly checks carried out in the flight simulator. In addition to the prescribed emergency exercises with carefully choreographed routines as part of standardized procedures, United Airlines[117] broadened the range of possible scenarios. This was known as Line Oriented Flight Training (LOFT).[118] During these training sessions, crews were confronted with minor technical issues, such as gradual loss of pressure due to a faulty valve, a blocked pitot tube causing incorrect speed and altitude readings, or a defective oil filter in a specific engine. However, rather than just observing and evaluating the crews' technical skills (e.g., their ability to identify and analyze a problem and act accordingly), the sessions also looked at how the individual crew members communicated with one another. The focus was no longer solely on standardized responses to emergencies, but also on the level of cooperation between those responsible for dealing with such situations. A discussion at the end of the session evaluated both these issues and allowed the participants to talk things over. Consequently, the captains realized that, although their sovereignty and dominance took some of the pressure off the rest of the crew, it also inhibited them to the extent that they felt unable to speak up. That meant any flow of information from crew to captain was a complete non-starter. In turn, the copilots became aware that, if they failed to report problems, their reserved and timid behavior toward their captains could cause tricky situations to become much worse. As it turned out, these practice hours in the familiar simulator environment were more effective than the previous seminar-room courses.

The second CRM generation: team approach

As time went on, it became clear that the definition of "team" used in Cockpit Resource Management was too narrow. The key factor here was an accident involving a DC-9 belonging to Air Canada (AC 797) in 1983. During a flight from Dallas to Montreal, a short circuit caused a fire to break out in the plane's rear washroom. The subsequent report by the NTSB[119] revealed

---

117 Simulators were becoming the norm in flight training from the start of the 1970s. They featured fully fitted cockpits, screens providing comprehensive visuals of the outside environment, and mobility in all three spatial axes. Today, pilots do their training exclusively in flight simulators as these are not only more cost-effective but also allow pilots to carry out critical maneuvers without any risk to safety.
118 For more detailed information, see Helmreich, R.L. and Foushee, H.C. (1993), pp. 28–30.
119 NTSB (1986).

that the cooperation in the cockpit had been fine, but that the cabin crew had not been properly integrated. This addressed another key problem. In terms of status, the cabin crew was even further down the ranks than the copilots and flight engineers. It was unthinkable for this "lowly" group of people to get involved in the affairs of captains, copilots, and flight engineers. With this in mind, let us take a closer look at the case of AC 797 (Figure 2.6).

The plane, a DC-9 assigned flight number AC 797, took off from Dallas at 5:25 p.m.[120] on June 2, 1983, with 41 passengers and five crew members on board. In the cockpit were captain Donald Cameron (51) and copilot Claude Ouimet (34), both experienced pilots with more than 10,000 flying hours each. They had been flying DC-9s for several years. The plane quickly reached its cruising altitude of 33,000 feet and everything seemed to be going normally. At 6:48 p.m., a crackling sound similar to an electrical discharge could be heard in the cockpit. Ouimet was eating his evening meal at the time. Neither he nor Cameron noticed the noise to begin with.

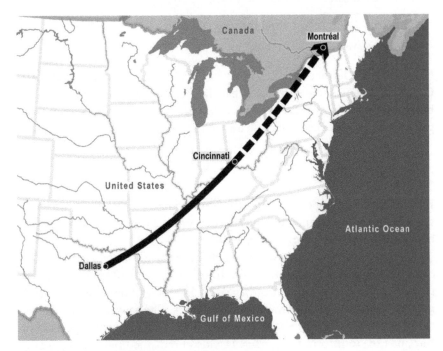

Figure 2.6 **Route of AC 797**

---

120 The investigation report uses Eastern Daylight Time. As Central Daylight Time applies in Dallas, the flight actually took off at 4:25 p.m. local time.

Cameron: "How is your seafood? Nice?"

Ouimet: "It's good."

Cameron: "Is the steak nice?"

Ouimet: "Different. A little bit dry, but okay."

At 6:51 p.m., after more crackling sounds, three circuit breakers popped out one after another.

Ouimet: "What was that?"

Cameron swore and tried to see which circuit breakers were affected. This was not an easy job, given that they were located on the wall right behind the captain's seat, but he soon managed to find the three in question. They were for the rear lavatory. Cameron pushed them back in. The crackling noise resumed and the circuit breakers popped straight back out again. Cameron tried again, but the result was the same.

Cameron: "Ha. Like a machine gun."

Ouimet: "Yeah, zap, zap, zap."

Cameron told Ouimet to put an entry in the technical log book for the Air Canada mechanics. He suspected that the flushing mechanism in the on-board toilet was jammed, causing its electric motor to overheat and popping the circuit breakers. Once again, he tried to push them back in.

Cameron: "Pops as I push it."

Ouimet: "Yeah, right."

By 7:00 p.m., both pilots thought the problem had been resolved because, apart from the fuses, they had received no other alerts. Cameron ordered his meal from purser Sergio Benetti (37).

At the same time, a passenger smelt burning toward the rear of the cabin. He notified flight attendant Judith Davidson (33). She guessed that the smell was coming from the rear lavatory, grabbed a fire extinguisher, and opened the lavatory door a crack. She saw thin gray smoke rising up, but no flames. She shut the door and asked her colleague Laura Kayama (28) to inform the purser. Davidson remained outside the wash-room. Kayama went to Benetti and explained that there was a fire in the rear washroom.[121] Benetti told her to inform the captain and then help

---

121 Davidson had in fact only mentioned smoke, but Kayama was sure she had also heard mention of flames and duly reported this to the cockpit. NTSB (1986), pp. 2, 70.

Davidson move the passengers from the rear of the cabin to the front. After that, the two women would be able to open the vents above the seats. Kayama went into the cockpit. Benetti took the nearest fire extinguisher and headed back to the rear of the cabin. He saw no flames in the washroom either, just black smoke that seemed to be coming from the wall cladding. He sprayed the washroom with the fire extinguisher and closed the door.

It was now 7:02 p.m. In the cockpit, Kayama spoke to the captain.

Kayama: "Excuse me, there's a fire in the washroom at the back. They [other flight attendants] went back to go to put it out."

Cameron: "Oh yeah."

Kayama: "They're still, well they're just gonna back now."

Ouimet: "Want me to go there?"

Cameron: "Yeah, go."

Ouimet: "The [circuit] breaker's fucked up."

Ouimet: "Got the breakers pulled?"

Cameron: "It's the motor."

Ouimet: "You got all the breakers pulled out?"

Cameron: "The breakers are all pulled, yeah."

As required in suspected fire situations, Cameron donned an oxygen mask. Ouimet went back to see what was happening. The smoke had become thicker and was already making it difficult to see in the rear part of the cabin. Ouimet returned to the cockpit to fetch some smoke goggles.

Ouimet: "I can't go back now, it [the smoke] is too heavy, I think we'd better go down."

The plane had just flown over Louisville International Airport in Kentucky, making it one option for the landing proposed by Ouimet. However, Benetti now returned and said the smoke had thinned.

Benetti: "I got all the passengers seated up front, you don't have to worry. I think it's gonna be easing up."

Ouimet: "Okay, it's starting to clear now."

Cameron: "Okay. Well I want – hold on then."

Ouimet: "I will go back [to see] if that appears better."

Cameron: "Yeah that's okay. Take the smoke mask."

A minute later, after Ouimet had returned to the cabin, Kayama reported to Cameron.

Kayama: "Captain, your first officer wanted me to tell you that Sergio has put a big discharge of CO_2 in the washroom, it seems to be subsiding, all right."

Cameron assumed that the smoke had been caused by a fire in the washroom waste bin and that Benetti had been successful in extinguishing it.[122] At that very moment, the first electrical systems failed and the master caution light came on.

Cameron [to air traffic control]: "Memphis Center [this was a mistake; he meant Indianapolis] this is Air Canada seven nine seven."

Indianapolis Center: "Canada seven nine seven, go ahead."

Cameron: "We've got an electrical problem here, we may be off communication shortly. Stand by."

By now, Benetti was back in the cockpit.

Benetti: "I was able to discharge half of the CO_2 inside the washroom even though I could not see the source. But it's definitely inside the lavatory."

Cameron: "Yeah, it's from the toilet, it's from the toilet."

Benetti: "CO_2, it was almost half of the bottle and it now almost cleared."

Cameron: "Okay, thank you."

Ouimet had now reached the rear washroom, but could no longer open the door as it had become too hot. He ordered Davidson and Kayama to keep the door shut, returned to the cockpit, and reported back to Cameron.

Ouimet: "I don't like what's happening. I think we better go down, okay?"

---

122 Cameron said this during the accident analysis. NTSB (1986), p. 3.

Cameron: "Okay."

Ouimet: "Okay, I'll be back there in a minute."

At this point, 7:07 p.m., the next electrical systems failed. The voice recorder stopped recording. One minute later, Ouimet sent a Mayday to Indianapolis Center and requested clearance to enter a descent. The controller in Indianapolis told the crew that they were 25 miles from Greater Cincinnati International Airport and asked if they could make it that far. Ouimet said yes.

During the descent, the smoke in the cabin intensified and gradually started to seep into the cockpit. Like Cameron, Ouimet now put on his oxygen mask. However, once they had the masks on, they could only read the instruments by leaning right forward. In that position, they were unable to see anything else apart from the instrument panel. Making things even more difficult was the fact that neither the elevator trim nor the heading indicator was working. The crew was only able to fly the plane by following the air traffic controller's instructions.[123]

Meanwhile, the cabin crew was distributing damp towels to the passengers, who were all gathered in the front section of the cabin. They also explained to the passengers how to open the emergency exits over the wings that the crew would be unable to open themselves, as they would be next to the main exits to open the front doors.

By now, several fire engines had taken up position along the planned landing runway, 27L, at the airport in Cincinnati. The controller asked Ouimet how many people were on board and how much fuel was left. All Ouimet said in response was: "We don't have time. It's getting worse here."

At an altitude of 3,000 feet, Cameron switched off the air conditioning system. He felt that it was making the fire and smoke worse. He and Ouimet opened the cockpit window a few times to improve visibility, but it was no use. New smoke kept billowing in. Despite this, at 7:20 p.m., Cameron managed to land AC 797 on runway 27L at Greater Cincinnati International Airport.

He switched off the engines immediately after landing. The flight attendants and passengers opened the two front doors and three of the four emergency exits over the wings. Eighteen passengers and three attendants were able to leave the plane via these exits before the sudden influx of

---

123 In this process the pilot does not receive the usual information regarding which course to follow, but is told to turn right or left. Once the plane has changed direction as intended by the controller, the crew is instructed to stop the turning flight.

Figure 2.7   **AC 797 on fire in Cincinnati**

oxygen caused the fire to explode through the cabin (Figure 2.7). Cameron and Ouimet had to leave the plane through the open cockpit windows. For 23 passengers, it was all too late. They either died of smoke inhalation or were unable to escape the flames. Almost an hour later, at 8:17 p.m., the firefighters reported that the fire was out. The subsequent investigation by the NTSB found that the fire had been caused by a short circuit in the motor of the electrical flush system in the rear washroom.

Among many other things, this case involves an interesting communication phenomenon that would also have occupied those responsible for developing the CRM concept. Basically, we tend to rate the value of information according to how we rate the person who delivers it to us. If it comes from someone we do not regard very highly, for whatever reason, this will have an effect on how we deal with the message they deliver. By the same token, we will listen to – and be inclined to act on – information from someone we respect or esteem. This phenomenon will not be new to any of us. However narrow-minded it may be, we, too, can unthinkingly accept information given to us purely because we like or respect the messenger.

The case of AC 797 is a fine example of this, and shows how we can fail to check something properly simply because the "wrong" person passed on the information to us. Take Kayama, for example, the flight attendant who apologizes before shyly informing the cockpit of the fire. "Excuse me, there's a fire in the washroom at the back. They went back to go to put it out." "Oh yeah," Cameron replies. Echoing cases we have looked at where younger copilots are scared of their superior officer, Kayama gets confused and cannot express herself clearly. "They are still, well they're just gonna back now." Now Ouimet, the person closest in rank to the captain, gets involved and asks, "Want me to go there?" Indifferently, Cameron answers, "Yeah, go."

Had Ouimet and Cameron taken the fire seriously at this point and realized the cause of it, they could have tried to land at the nearby airport in Louisville, Kentucky. Instead, they and the rest of the crew were disorganized and confused. Wanting to check out what is going on, Ouimet heads off to the rear of the cabin without putting on smoke goggles.[124] Once at the scene, he realizes he needs a pair, and returns to fetch them before heading back to the washroom. This wastes crucial minutes. During this time, purser Benetti establishes that the smoke is coming from the wall cladding of the washroom and decides to spray half a canister of extinguisher on it instead of using the axe to first open the wall. Meanwhile, the captain has made the assumption that the fire is in the washroom waste bin, even though he also touches on the idea that the problem could be the motor in the toilet flush system. Neither notion concerns him very much and he continues with the flight. There is no discussion between Cameron and Benetti. In the meantime, Ouimet suggests that perhaps they should land at the airport in Louisville, Kentucky, after all. However, he does not pursue this idea further once Benetti says the smoke is not as bad as it was. Due to the general disorder in the whole communications process, no one realizes that the smoke had thinned only because Benetti had sprayed half a canister of extinguisher in the washroom. By the way, the Air Line Pilots Association (ALPA) as well as Ouimet stressed that Cameron did not show neglect by deciding to continue the flight instead of immediately diverting to Louisville and therefore they rejected the NTSB's conclusion. Both cite the ambiguous information from the cabin crew as a reason for their decision to continue the flight. But should the cabin crew have expressed themselves more clearly? They probably would have, if the hierarchical difference had not been so great.

In sum, neither the cockpit nor cabin crew were able to communicate and work together properly in a gradually worsening emergency situation. According to the NTSB investigation report, had they done so, they would definitely have managed to land in Louisville rather than Cincinnati. This would have reduced the flight time by at least five minutes. Those few minutes might have been crucial, as the thickening smoke in the cabin killed many, if not most, of the passengers who died.

---

124 Ouimet had given his goggles to the captain, who had difficulty retrieving his own.

During the next workshop organized by NASA in May 1986, the term "Crew Resource Management" was coined for the first time and would later become established in the industry. NASA was keen to collate reports from airlines such as United, Delta, American Airlines, British Airways, and Lufthansa, as well as the US Air Force and the aviation safety authorities.[125] By this point, each of these institutions had an officer responsible for ensuring that crews received "human factor training" and for reporting on experiences from training environments and everyday flight situations. Due to the growing volume of research on team behavior, virtually all the major international airlines had by now developed and implemented Cockpit Resource Management training, even though they were not legally required to do so at that time.

One of the most interesting pieces of research presented at this second major workshop came from a NASA project carried out by Foushee, Lauber, Baetge, and Acomb. Initiated by the US Congress, this study on "Crew Performance as a Function of Exposure to High-density, Short-haul Duty Cycles"[126] aimed to identify the effect that fatigue has on the performance of plane crews. It also examined the extent to which the long hours that the crews were now expected to work – following the deregulation of the aviation industry at the end of the 1970s – might impact on flight safety. The crews participating in the study undertook several flights in a Boeing 737 simulator. In addition to normal flight conditions, these simulations also involved dealing with technical problems. A comparison was made between two different groups. On one side were crews that had completed a long day of flying together and were deemed to be fatigued; on the other were crews that had been newly formed and came to the simulator flight rested and ready. Both groups tackled the same program.

As expected, after the simulator flight, the fatigued pilots complained that completing the tasks had drained their energy. However, what took everyone by surprise was the fact that they actually performed better than the rested cockpit crews that had just been put together for the first time. Despite their exhaustion, the crews that had already worked together made fewer errors than the others. In this respect, this study confirmed the importance of the team concept within a group. Still this finding could not be fully translated into the practical airline environment; it would have entailed a massive increase in the personnel required. (If one member of the team fell ill, it would mean having to find a complete replacement crew.) Things are different in the military aviation sector, where crews fly together for longer

125 Orlady, H.W. and Foushee, H.C. (1987).
126 Foushee, H.C. et al. (1986).

periods, sometimes for more than a year. In this sector, the level of performance required in battle situations takes priority over cost-efficiency.

Nonetheless, the abovementioned study did have its effect. In 1986, the FAA drew up a regulation that would make the CRM concept an integral part of the training process for commercial pilots in the United States. After its implementation in 1989,[127] cockpit crews were no longer trained separately, but worked as a team from the very start.[128] As the years passed, CRM training became generally accepted. This was due partly to the NASA studies made available to all, and partly to accident reports on various crashes, some of which we looked at in the first section of this text. Only the psychological elements of the training continued to encounter resistance, with pilots occasionally referring to them as "psycho-babble."[129]

The fundamental aim of CRM training was – and is – to shape or correct cockpit crews' attitude. In his excellent overview of the various elements of CRM, Tony Kern (working with Jack Barker) described the pilot's basic attitude as follows: "Pilots come to aviation from across the spectrum of society, and we fly for a wide variety of reasons. Some motivations lend themselves to a solid base of professionalism, and others are potentially hazardous."[130] Based on this, he says CRM aims "to identify hazardous attitudes and [...] develop tools to recognize, avoid, mitigate, or modify these attitudes when we encounter them."[131] Drawing on their extensive flight experience, Kern and Barker identified the most hazardous mindsets.

Top of the list is one that we are probably all familiar with. Kern and Barker refer to this as *pressing*, meaning a pilot's all-consuming desire to "get the job done," combined with the conviction "I can do it." In the flight environment, this type of attitude can mean getting too close to a weather front; trying to cover a certain distance without refueling as planned, even though the decision is risky; wanting to check for themselves whether an area of bad weather is indeed as bad as air traffic control center reported; feeling like the master of the plane, including its passengers and machinery. No one can stop them – they know best how to get things done. This attitude almost automatically includes the macho persona or, in our case, the macho pilot, whom Kern and Barker describe as someone who tends to "see the sky as their own personal playground where they can prove their prowess to lesser mortals. [...] the macho pilot is often mostly boast, which can place

---

127 FAA Advisory Circular AC, pp. 120–151.
128 Hackman, J.R. (1986), p. 23.
129 Helmreich, R.L. and Foushee, H.C. (1993), p. 3.
130 Kern, T. (2001), pp. 122.
131 Ibid.

them in a position of having to attempt a maneuver merely to prove – either to themselves or others – that they can actually do what they claim."[132]

This goes hand in hand with the feeling of being *untouchable*, that is, being convinced that accidents only ever happen to other people, while you yourself are immune to everything. For a pilot who feels he is untouchable, escaping a dangerous situation by the skin of his teeth will simply serve to confirm this conviction. He will never see it as a warning.

Such feelings of omnipotence also lead to *impulsiveness* and all the risks that it entails. "As controllers," write Kern and Barker, "we want to take charge, and this often leads to situations where we act impetuously. This is especially true when we are confronted with situations that have no clear-cut answer. We want to make a decision and we want to make it NOW, and yet we may not have all the required information to make the call."[133]

At the opposite end of the scale from this exaggerated self-confidence lies the equally dangerous state of *resignation*, or feeling that "there's no point anyway." In crisis situations in the air, this is a sure-fire way to sign your own death warrant. "The nature of the situation," write Kern and Barker, "will dictate what options are available, but giving up is not one of them."[134] We found this fatal resignation in first officer Meurs and in flight engineer Moyano. (For comparison, in part III, we will look at one of the best flight examples of people not giving in and fighting against all odds.)

A close relation of resignation is *indifference*,[135] a state that can prove particularly dangerous for experienced pilots because it takes them one step closer to inattentiveness. Captain Marsh of JAL flight 8054 definitely reached an extreme when he thought he could still fly his cargo plane even though he was drunk. However, such thoughts are undoubtedly underpinned by the conviction that every flight is routine and nothing will ever happen. It is a dangerous state of mind that the pilot suffering from it must learn to overcome. The following quote comes from John W. Olcott, former President of the National Business Aviation Association: "Fundamentally airmanship is about acceptance of responsibility for everything that involves your aircraft – its position, its equipment, its crew, and the instructions you choose to follow [...] from air traffic controllers. [...] While successful airmanship requires using knowledgeable input from all available resources, nothing allows abdication of responsibility. Maintaining such vigilance in an atmosphere that so easily nurtures complacency is an awesome challenge."[136]

---

132 Ibid, pp. 129–130.
133 Ibid., p. 133.
134 Ibid., p. 136.
135 Ibid., p. 138.
136 Ibid., p. 140.

In part III we will find a practical example of what Kern and Barker term *air show syndrome*. It involves pilots flying in such dangerous, foolhardy, and daring ways that it leaves people speechless – despite or because of the mortal dangers involved. For these men, rules are for sissies only. The cure, if there is one, means steering them toward recognizing their motives and overcoming their need for admiration with professional help and coaching. If a pilot refuses to do that, or if it fails to have the desired effect, the only option left is to impose sanctions.

Kern and Barker also identified the danger of *emotional jet lag*. This refers to the difficulty individuals have in coming to terms with mistakes they have made – an attitude that is a frequent consequence of the high-performance qualities pilots have to have. Even in cultures where errors are permitted and accepted, people making them will, in a first impulse, be annoyed with themselves and prefer someone else to have been responsible. For the perfectionist, making a mistake is particularly insufferable. If they admit an error – a major achievement in itself – it will still gnaw at them for a long time. As Kern and Barker rightly point out, in the air there is no time to waste brooding over mistakes. "You cannot allow your mind to get behind the aircraft. If you do, much bigger – and perhaps deadlier – mistakes are likely to follow."[137] While flying, or working in any situation demanding our full attention, there is no time for dwelling on mistakes, but only time to focus on the next step and push the ignominy to the back of our mind. Once the flight is over, people can use a debriefing to analyze their error with their colleagues from the cockpit and/or cabin, but not dwell on it while all senses are needed.

Third and fourth generation: integrated team approach

At the start of the 1990s, the FAA initiated further expansions to CRM training. This time the spotlight also fell on areas like flight operations, which covers the planning and supervision of flights on the ground, including maintenance. In the beginning, the individual groups – cockpit, cabin, operations – were taught separately. By the end of the 1980s, it had become clear that it made more sense to put these groups through CRM training together. As the major airlines in the United States and Europe had begun to take a more systemic approach, key factors such as corporate culture were also taken into account.[138] Ideally, it was meant to lead to organizational

---

137 Ibid., p. 144.
138 Pariès, J. and Amalberti, R. (1995).

resource management[139] and/or company resource management.[140] So far, the latter has not come to pass. Currently, CRM training still remains limited to pilots, flight engineers, flight attendants, and ground staff.[141]

Yet in the time since the first NTSB/NASA workshop in 1979, a new generation of captains have gradually emerged. They have come to know and value CRM training. For them, the concept is no longer associated with a loss of authority, especially as stripping them of power had never been part of CRM. Their CRM motto is "authority with participation," whereas the rest of the crew follows the motto "assertiveness with respect."

By about 1990, some 10 years after CRM's first tentative steps, it had turned into an accepted part of cockpit and cabin crew training.[142]

Fifth generation: error management

Based on research carried out by James Reason, a professor at Manchester University,[143] CRM has been viewed largely as a process of error management since 1997. Drawing on numerous case studies, Reason demonstrated that it is impossible to prevent human error, even with enormous technical outlays and multiple monitoring systems. This is particularly true in complex risk areas such as aviation, the chemical industry, and medicine. By means of illustration, he compared the safety mechanisms that are used to Swiss cheese and described the chain of errors as holes of varying sizes. Even the very smallest of these holes penetrate through all the layers[144] (Figure 2.8).

Reason's main conclusion was that, although it is impossible to eliminate human fallibility, it is possible to reduce the error rate by adjusting the conditions under which people work. For Reason, it is essential to identify and correct errors and, ideally, learn from them.[145] According to Robert Helmreich, when errors are fundamentally unavoidable, CRM becomes a corrective process with three lines of defense,[146] namely, *preventing errors* (using training, process standards, checklists), *identifying errors* (using reciprocal checks, or cross-checks, between crew members), and *addressing*

---

139 Heinzer, T.H. (1993).
140 Pariès, J. and Amalberti, R. (1995).
141 Helmreich, R.L. and Foushee, H.C. (1993), p. 3.
142 Helmreich, R.L. et al. (1999), p. 6.
143 Reason, J. (1990, 1997).
144 Reason, J. (1997), p. 12.
145 Ibid., p. 25.
146 Helmreich, R.L. et al. (1999), p. 7.

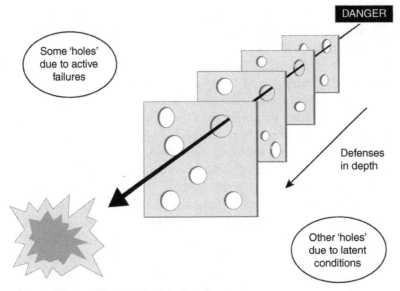

Figure 2.8 **Swiss Cheese Model after Reason**

errors (regardless of the hierarchical position of the crew member). Errors are discussed in a debriefing, without any blame being apportioned.

To make the outcomes of these critiques available to other crews, most airlines have introduced an anonymous reporting system that enables crew members to provide information about errors that occur during flights. In addition, the NASA-run Aviation Safety Reporting System (ASRS) has been in place in the United States since 1976 and provides a way to report errors anonymously. Other countries have similar systems, all of which are combined in the International Confidential Aviation Safety Systems.

CRM training

CRM training, including annual refresher courses, has been obligatory for all flight crews worldwide since 1997 – not least because of the Birgenair crash described in part I.[147] The CRM training now conducted by the major airlines[148] mainly focuses on developing skills such as understanding

---

147 JAR-OPS 1 Subpart N. An outline of the CRM training requirements has been detailed by the Civil Aviation Authority – Safety Regulation Group (2006).
148 A good summary of the key points is available in the overview by the UK Civil Aviation Authority – Safety Regulation Group (2006).

situations, planning, decision making, and communication. These are usually illustrated using case studies that are often based on plane crash investigation reports. We have seen some examples in part I. This is accompanied by role plays to train and practice fact-based communication techniques. All the training and exercises are tailored to the participants' specific flight environments. The psychological role plays applied in the first years of CRM are no longer used today, nor are the rigid behavioral models originally favored in the early days. To put it in a nutshell, the aim is to motivate plane crews to embrace an open culture of errors.

Elements of CRM training such as communication and decision making are also used in management training. However, unlike corporate seminars, CRM training does not focus on strategies for business success. Instead, it analyzes accident case studies. Even so, CRM training alone would not be enough to teach people how to deal with errors openly. For that, we need a culture in which encouraging people to report errors is paid more than just lip service. The most important factor is to deal with errors without imposing sanctions. Without that, we might as well forget about the whole idea.

When it comes to decision making, CRM training features a framework that is now used all across the world, that is, the FORDEC model developed by the German Center for Aeronautics and Astronautics (DLR). Each letter of the acronym represents one step in the decision-making process: analysis of all relevant *facts*, identification of possible *options*, the *risks and benefits* of the decision, followed by the *decision* itself, and its *execution*. In this respect, this is similar to the decision-making processes in normal day-to-day business scenarios. The only feature particular to the aviation sector is the *cross-check*, or continuous monitoring of the decisions taken. A practical example will help to illustrate this. Imagine a hydraulic system has failed. For the cockpit, that means the crew members gather all the facts, discuss the options available to resolve the problem, decide which one to apply, constantly monitor the consequences of this decision, and make adjustments when necessary.

If we look at today's cockpit crews[149] in action, we can see two alternating forms of communication. One takes place with reference to the flight, mostly in the form of routine "checking" calling for standardized notifications and comments. The moment a scenario is no longer part of this routine, pilots switch to a different mode. The dialogues become more

---

149 As part of a research project carried out over several years, the author observed numerous crews from a major European airline in normal flight conditions and during training sessions in a flight simulator.

personal and – depending on the danger of a situation – increasingly tense. Internal agitation may find release through swearing. The main thing, though, is that everybody talks freely and without fear of the others.

Generally, cockpit members check and discuss their analyses, assessments, and thoughts with their colleagues before the flight with an initial briefing or during the flight preparations. Possibly at that point they all meet for the first time. They discuss the key factors of the flight, such as weather, traffic at the departure and destination airports, any particular issues concerning passengers or cargo, and the order in which captain and copilot will switch roles in flying the plane. The briefing lasts between 15 and 20 minutes for short-haul flights, and 30 to 45 minutes for long-haul ones. A study of US airlines[150] found that it determines the atmosphere for the subsequent flight: if the captain acts in a dominant fashion, the other members of the crew will become cautious and reserved; if he treats them as colleagues, they will do their best to cooperate with him. In conflict situations, the captain has the final say. He is – and remains – the pilot in charge. A joint debriefing takes place after the last flight. The key here is that the captain lets his crew members speak first to express their feelings about what was good or bad about the flight, without accusations and self-criticism, but just reflections on the decisions made and actions taken. If something went wrong, the question is only, why? If the crew realize errors later, they can report them anonymously to the manager of flight operations or, as mentioned previously, pass the information on through the error recording systems; we shall come to this again in part III. Often, such information will feed into CRM training.

Although the power status and responsibilities borne by captains have barely changed in the last 50 years, the gulf between the captain and the rest of the crew has grown considerably smaller since the implementation of CRM. If asked, the majority of captains today will say that they do not view the open communication they now have with other crew members as a disadvantage, but as a relief. In short, the CRM concept was initially seen as a way to prevent accidents, or at least to prevent them as far as possible. Eventually, it resulted in changes in behavior and improved team performance. In some cases, the improvements exceeded expectations. We will come to that when dealing with our cases in part III.

---

150 Ginnett, R.C. (1987).

Part III Post Crew Resource Management

UAL flight 811: we got a control problem here

The crew of United Airlines flight 811 (UAL 811) paved the way. Ten years after the introduction of Crew Resource Management, they were the first to *publicly* speak out in favor of the concept after being motivated by the following incident: on February 24, 1989, they were en route from Honolulu to Sydney in a Boeing 747–122 when, 17 minutes after takeoff, the front right cargo door sprang open. This caused an explosive decompression that ripped off the door, parts of the fuselage on the right of the plane, and sections of the cabin. Some of the debris entered both of the engines on the right, causing them to fail. Even in these circumstances, the crew managed to keep the heavily loaded plane under control and make it back to the airport in Honolulu.[151] Much of this extraordinary achievement they attributed to their CRM training. Accordingly, we will see how strikingly different their interactions with one another were compared to the crews in our previous cases.

After UAL flight 173 in 1979 – we dealt with in it part I – United Airlines hit the headlines because both the first and second officers had been unable to make their captain aware of the decreasing fuel reserves clearly and promptly. At that time, United Airlines was one of the big-name airlines in the United States. The fact that calamitous and fundamentally avoidable errors had occurred during one of its flights did not fit the airline's image of itself. Even worse was the possibility of the same thing happening again. As a result, United was one of the main airlines that actively participated in the NTSB/NASA workshops and made a real effort to train

---

151 All the following information is based on the NTSB investigation report. Cf. NTSB (1992).

cockpit crews based on CRM criteria. As early as 1980, they had been sending their crew members to at least one CRM training course per year. Just the same, it took nine years before a crew like that of UAL 811 spoke out so unambiguously in favor of CRM training.

Let us get back to our case. The captain of flight UAL 811 was David Cronin (59). He was due to retire very soon and was on one of his last flights.[152] He had been flying as a United captain for 35 years and had completed a total of 28,000 flying hours. He had been in charge of Boeing 747s for the last four years.

Copilot Gregory "Al" Slader (48) had been flying for United for 25 years. He, too, was a very experienced pilot with more than 14,000 flying hours. He had been qualified to fly the Boeing 747 for two years but, with only 300 hours flying time, his experience with this type of plane was still limited.

Flight engineer Randal Thomas (46) was only slightly younger and had been working for United for 20 years. Of his nearly 20,000 hours in the air, 1,200 had been spent in the Boeing 747, which he had been flying just short of two years.

At the time of the accident, the Boeing 747–122 was 20 years old (Figure 3.1). It had completed more than 15,000 rotations and notched up

Figure 3.1   **Boeing 747–122 on takeoff. The cabin is below the cockpit**

---

152 The upper age limit for pilots in the commercial sector in the United States at that time was 60.

a total flying time of almost **59,000** hours. Although the plane had been maintained as required by regulations, the various crews who had flown it in the 10 weeks prior to the accident had made 11 entries in the technical log book indicating problems with the front cargo door. Most of them referred to difficulties in locking it. Ultimately, however, it was still functional, and the laborious error analysis process had been scheduled for a later date.

Shortly after midnight[153] on February 24, captain Cronin and his crew were preparing to fly from Honolulu to Sydney (Figure 3.2). They had already been working together for several days and had just had a day and a half of rest. There were 15 cabin crew members and 337 passengers on board. The plane's takeoff weight of **698,000** pounds was just below the maximum permitted limit of **706,000** pounds. The Boeing was pushed back from the gate after a slight delay of three minutes. Nineteen minutes later, the plane took off into the night. As Cronin saw thunderstorm cells on the weather radar, he decided to leave the seatbelt sign on during the ascent.

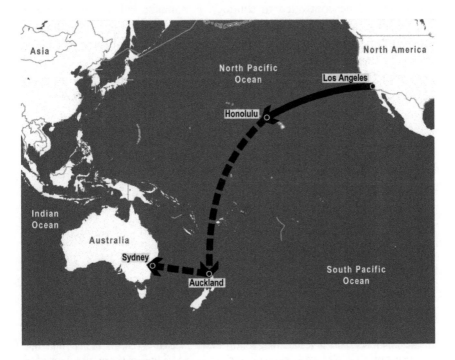

Figure 3.2   **Route of UAL 811**

---

153 All times are Honolulu local time.

At 2:09 a.m., 17 minutes after takeoff, at an altitude of 23,000 feet and a location 70 miles southwest of Honolulu, the crew heard a muffled noise. A powerful shudder went through the plane, followed by something that sounded like a "tremendous explosion."[154]

Cronin: "What the fuck was that?"[155]

Slader: "I don't know. It looks like we've lost number three engine."[156]

In the cabin below, chaos broke out.[157] Like a bolt from the blue, a hole had been ripped into the cabin wall. A deafening, icy cold wind immediately swept through the cabin, enveloping everything in fog. Objects and scraps of paper went flying through the air, and the oxygen masks fell from the ceiling. Where the hole gaped, an entire row of seats was missing, along with its passengers.

Fog also started to form in the cockpit above, and visual and acoustic warning signals were activated. Everything indicated that an engine had failed. It was also obvious that something had happened down in the cabin. The plane continued to shudder. For maybe half a minute, there was no information on the voice recorder.[158] The recordings started again once captain Cronin began the rapid descent to reach an altitude of 10,000 feet or below, where the oxygen level would be sufficient. Slader informed Honolulu Center air traffic control of the emergency. The following parts of the conversation are extracts of the communication in the cockpit.

Slader: "United eight eleven heavy, we're doing an emergency descent."

Slader [to Cronin]: "Put your mask on, Dave."[159]

Slader [to Thomas]: "Go through the procedure for number three [engine]. I think we blew a door or something."

Cronin [to Thomas]: "Tell the PA [public address] – to get prepared for an evacuation. We don't have any fire indications?"

---

154 NTSB (1992), p. 2.
155 Transcript Cockpit Voice Recorders UAL flight 811, N4713U (1990).
156 Number three engine is the inside right engine of the Boeing 747.
157 Rutherford, M. (undated).
158 NTSB (1992), p. 25.
159 Cronin was so busy in maintaining control of the crippled 747 that he did not find time to put on his oxygen mask. However, the supply and fill lines of the oxygen system had been broken as a result of the blown cargo door. Therefore, neither crew nor passengers received any oxygen despite wearing their masks. Due to Cronin's immediate descent and the fact that the aircraft had not reached its cruising altitude, all on board were able to survive the initial stage of the crisis without supplementary oxygen.

Thomas: "No, I don't have anything."

Slader [to Honolulu Center]: "Center, United eight eleven, you want to have the [fire] equipment standing by. Company [United Airlines] notify please."

It was now 2:12 a.m. Slader and Thomas went through the checklist for shutting down the engine. At this point, no one in the cockpit knew exactly what had happened. With the alerts and warnings still sounding in the cockpit, there was no time to go down into the cabin, find out what was going on, and speak to the cabin crew. Cronin kept trying to stabilize the erratic Boeing 747. Slader was speaking to air traffic control, who alerted the US Coast Guard. They were to stand by in case the plane had to make an emergency landing on the water.

Thomas: "Lots of fuel. Should we dump [fuel]?"

Slader: "Honolulu, United eight eleven heavy, we're gonna level at nine thousand [feet] while we assess our problem. We're coming back direct."

Honolulu Center: "United eight eleven heavy, roger. Keep the center advised."

Cronin: "We got a control problem here."

An announcement from the cabin was now audible in the cockpit, delivered via a megaphone rather than the normal on-board system: "Everyone take your seats – everyone take your seats."

Slader [to Thomas]: "Start dumpin' the fuel."

Thomas: "I am dumpin'."

Honolulu Center: "United eight eleven heavy, when able, forward the souls on board and fuel at landing."

Slader: "Okay, stand by. We'll give it to you as quickly as possible."

Cronin [to Thomas]: "What are you dumping down to? We've got a fucking control problem here."

Thomas: "I'm dumping everything."

Cronin: "We got a problem with number four engine."[160]

Thomas: "Yeah, number four looks like it was out too."

---

160 Outside engine on the right-hand side.

Slader: "Well we got EGT,[161] we got N1."[162]

Cronin: "Okay, we got a problem with number four engine, too."

Slader: "Can you maintain two forty [knots]?"[163]

Cronin: "Yeah, just barely."

Slader: "But we're losing altitude."

Cronin: "I know it."

Slader: "Center, United eight eleven heavy – do you have a [radar] fix on us?"

Honolulu Center: "Affirmative sir, I have you on radar."

Slader: "Okay, it appears that we've lost number three engine and we lost – we're not getting full power out of number four. We're not able to hold an altitude right now. We're dumping fuel."

Honolulu Center: "United eight eleven heavy, roger. I show you six zero miles south of Honolulu at this time."

Thomas: "I haven't yet talked to anybody yet. I couldn't get to them. You want me to go down the stairs and check?"

Cronin: "Yeah, let's see what's happening down there. I can't hold altitude."

Thomas left the cockpit and went down to the cabin.

Cronin: "We're getting more rumble. No fuel flow on number four engine."

Slader: "How can we have no fuel flow if we got N1 and EGT?"

Cronin: "We must be losing fuel like mad."

Slader: "Watch your heading, watch your heading. You want to go direct to Honolulu."

Cronin: "Yeah."

Slader: "What a fuck of a thing to happen on your second to last month."

Slader noticed the fire warning on the instrument panel above them. "You got a fire out there."

---

161 EGT: Exhaust gas temperature of the engine.
162 N1: Speed of the engine's low-pressure rotor.
163 Approximately 445 km/h.

Cronin: "There's a fire out there?"

Slader: "Yeah, it looks like it's engine number four."

Cronin: "Go through the procedure to shut down the engine."

Slader: "We're not gonna be able to hold this altitude on two [engines]."

Slader and Cronin went through the engine-fire checklist and shut down the engine.

Honolulu Center: "United eight eleven heavy, pilot's discretion[164] descend to four thousand."

Cronin: "Okay. Four thousand. We got a fire on the right side. We're on two engines now."

Thomas returned to the cockpit at 2:17 a.m. "The whole right side – the right side is gone. From about the one right back it's just open. You're just looking outside."

Cronin: "What do you mean?"

Thomas: "Looks like a bomb. The fuselage – it's just open."

Cronin: "Okay, it looks like we got a bomb that went off on the right side. All the right side is gone."

Thomas: "Some people are probably gone. I don't know."

Cronin: "We got a real problem here."

Slader [to Honolulu Center]: "Center, United eight eleven heavy, now you've got to give us a vector direct Honolulu. We're losing the VOR."

Thomas [to Cronin]: "Zero three zero. Can you maintain heading now and altitude?"

Cronin: "Not really. We shut down number four. We're on two engines."

Slader [to Honolulu Center]: "We're down to sixty five hundred (feet) and we look like we can hold this altitude. We evidently had a bomb or something. A big section of the right side of the aircraft is missing."

Honolulu Center: "United eight eleven, you're missing the right side of the cabin or the right wing, sir?"

---

164 Pilot's discretion: clearance that can be implemented based on the pilot's judgment.

Slader: "That's affirmative. We're missing a section of the right side of the airplane. Part of the fuselage is missing and we've lost engine number three. We've got engine number four shut down. It appeared as if we had a fire out there. We want all medical equipment we can get and all the equipment we can get standing by."

Honolulu Center: "United eight eleven heavy search and rescue has launched a helicopter to intercept and aid you in returning back to Honolulu."

Slader: "Roger understand."

It was now 2:19 a.m.

Cronin: "Okay, what's our stall speed?"[165]

Slader: "I wouldn't go below two forty [knots]."

Cronin was aware he was treading a very fine line. If he flew slower than 240 knots without the flaps extended, he risked going into a stall and crashing. If he went faster than 250 knots, the plane could be ripped apart. In other words, with an almost fully-loaded Boeing 747 and only two engines, he had a speed tolerance of just 10 knots, or 19 km/h. Keeping within that limit demanded concentration and an extremely delicate touch.

Slader [to Honolulu Center]: "Honolulu, United eight eleven, we do plan to evacuate on the runway."

Honolulu Center: "United eight eleven heavy roger, land at your choice of runway, sir."

Thomas tried to inform the cabin of the evacuation, but realized that the on-board telephone system was no longer working.

Honolulu Center: "United eight eleven heavy, Honolulu wind at the airport at zero six zero at one zero."

Cronin: "What's the longest runway?"

Slader: "Eight right, I believe it is."

Cronin: "What? Ask him what the ... "

Slader: "Watch your altitude."

Cronin: "Yeah. We're going to four thousand, right?"

---

165 The stall speed corresponds to the minimum speed.

Honolulu Center: "United eight eleven heavy, contact Honolulu Approach now, one one eight point one, if able."

Slader: "Honolulu Approach, United eight eleven heavy is with you."

Honolulu Approach: "United eight eleven, runway four right. You're cleared to land."

Cronin: "Is that the longest runway?"

Slader [to Honolulu Approach]: "Okay we need a long runway. Eight is longer, isn't it?"

Honolulu Approach: "You can have eight left."

Cronin: "Eight left – okay. You want to set me up on that."

Slader: "Watch your airspeed."

Cronin: "I got max [power] here. I don't know if we're gonna make this. I can't hold altitude."

Slader: "Okay. Don't – okay, we have twenty-four miles to go and we're drifting down slowly."

Thomas: "You're gonna make this!"

Cronin: "Huh?"

Thomas: "You're gonna make this!"

Slader: "Make sure we don't hit any fuckin' hills on the way."

Cronin: "I need a long final [approach]. Tell him I need a long final."

Slader: "You can see in a minute. We're at four [thousand feet]. We're twenty-one miles out. We're in good shape, we're in good shape. Okay, now what do we want to do about the gear?"

Cronin: "We're gonna hold that until we get on the glide slope. What we're gonna do is go on a two engine approach. You want to read me a checklist."

Thomas: "Yeah, I got it out – when you're ready."

Honolulu Approach: "United eight eleven, I need souls on board if you have it."

Slader to Honolulu Approach: "We're too busy right now. It's two hundred and something." [As stated earlier, it was in fact more than 300.]

Honolulu Approach: "Okay."

Slader: "We're seventeen miles out, Dave."

Cronin: "What's the minimum speed right now? What have we got on fuel weight? Six ten [thousand pounds] – oh, fuck."

Thomas: "Don't get any lower, captain."

Cronin: "I know. Let's try one degree [of flaps] and see what happens."

The Boeing still weighed more than **600,000** pounds. Even though more fuel was being offloaded all the time, the plane would exceed the normal landing weight. The cockpit crew had no idea whether they would be able to deploy the landing flaps. Cronin requested the longest runway because they would be landing heavy and fast, meaning a long roll-out after touch down.

Slader [to Thomas]: "Watch your hydraulics."

Thomas: "Alright."

Slader [to Cronin]: "How do your controls feel?"

Cronin: "Alright so far."

Cronin: "Okay, I got to slow it down a little bit."

Thomas: "Do not go below two ten [knots] though. Are you gonna try to evacuate the airplane after landing?"

Cronin: "You bet."

Cronin: "Where's the airport?"

Honolulu Approach: "Okay, turn another fifteen degrees left, United eight eleven heavy."

Slader: "Okay, we've got the island in sight and we show we're only eleven miles from the airport, but we don't have the airport."

Honolulu Approach: "Okay, we'll turn the lights on high."

Thomas: "Can you make the turn here?"

Cronin: "Yes."

Slader: "We're higher than hell."

Slader: "You got the airport?"

Cronin: "No."

Thomas: "It's right over here to your right."

Cronin: "Okay. Let's try."

Slader [to Honolulu Approach]: "Are we good for terrain [clearance] out here?"

Honolulu Approach: "Yes you are."

Slader [to Honolulu Approach]: "Okay, we've the airport now, United eight eleven heavy."

Honolulu Approach: "Eight eleven is cleared to land eight left. Equipment standing by."

Thomas: "I hear people screaming back there."

Slader: "She [the purser] is yelling at them to sit down."

With a gaping hole in the cabin, some of the passengers were now beside themselves with panic. Once again, the purser had to resort to a megaphone to prepare for the evacuation.

Slader: "Are we depressurized?"

Thomas: "Yes. Have about two minutes until we touch down."

Cronin [to Thomas]: "How are you doing?"

Thomas: "I'm fine. I'm trying to catch up."

Slader [to Honolulu Approach]: "United eight eleven heavy, can you turn down the strobe [light] a little bit?"

Honolulu Approach: "Alright."

Cronin: "Okay. Final check."

Thomas: "One hundred feet. Fifty feet."

Cronin: "Center the trim, center the trim."

Thomas: "Thirty [feet above the runway] – ten – zero."

Slader: "We're on."

Thomas: "Shut down the engines."

Slader: "Okay. We're stopping here, United eight eleven heavy. We are evacuating the airplane."

Honolulu Approach: "United eight eleven, roger. You got the airport."

The Boeing came to a stop in the final third of runway 08L (Figure 3.3). The evacuation of the plane went without a hitch. However, nine business class passengers had been flung out over the Pacific through the hole in the cabin. Despite intensive searches by the US Coast Guard, they were never found. All the other 328 passengers and the 18 crew members survived the accident virtually unscathed.

Figure 3.3   **United Airlines 811 after landing**

The NTSB quickly identified the cause of the accident: the front right cargo hatch had sprung open during the ascent at an altitude of around 22,000 feet,[166] ripping open parts of the fuselage. The damaged exterior was unable to withstand the cabin pressure[167] and more sections were torn off, leaving a 10 by 15 feet hole in the cabin. During a laborious search operation, the US Navy found parts of the cargo door on the sea bed. These were recovered from a depth of 14,200 feet[168] at the end of September 1990.[169] Subsequent analysis of the door revealed design problems with the locking mechanism. In the wake of this incident, Boeing modified the cargo doors on all its 747s.

The behavior of the crew could not be criticized in any way. Both the cabin crew and cockpit team had done an amazing job. The cabin crew had excelled themselves: overcoming the shock of seeing passengers sucked out through the hole, calming the other panicking passengers, preparing for an emergency landing, and finally instigating and managing the evacuation.

---

166 6,600 meters.

167 During commercial flights, cabin pressure is regulated in such a way that a maximum altitude of 10,000 feet is not exceeded (normally 5,000 to 7,000 feet). At cruising altitude, the cabin of a plane is therefore exposed to overpressure.

168 Approximately 4,200 meters.

169 See NTSB (1992), pp. 26–27.

Cronin, Slader, and Thomas also had their hands full: not knowing the full extent of the damage caused by the explosion; the shuddering plane; visual and acoustic warnings; the engines failing one after another; making it almost impossible to maintain altitude; the subsequent descent; and the emergency landing. Altogether it was an impressive catalog of stress factors[170] that they successfully dealt with. Nonetheless, we can assume that the necessary measures – and the ability to stay calm in an emergency – had been practiced previously in a simulator. Most of the captains we encountered in part I would have known how to deal with a situation like this, at least in theory; even so, it is worth considering if they would have managed to land UAL 811 safely, given their lack of cooperation with their copilots and flight engineers.

So, the main difference between this case and the previous cases is the cohesion of the cockpit crew. Cronin focused on the job of flying the plane in risky conditions. Slader and Thomas took some of the pressure off him, with Slader taking care of communications with air traffic control in Honolulu. Naturally, the age and experience of the first and second officers played a role here. Still, we may remember that these factors were similar in the cases of KLM 4805 and UAL 173, where the crews had no influence on how events developed. The final decisions were made by captain Cronin, but Slader and Thomas had enough freedom to act on their own initiative, too. Thomas was perhaps marginally less integrated than Slader, but that was due in part to the way the tasks were distributed. In the first instance, the flight engineer's job was – and is – to monitor the operation of the flight, not actually fly the plane. They are trained primarily as engineers, sit in a right angle behind the pilots and, rather than looking ahead as the pilots do, their eyes are fixed on their own console. Nevertheless, they are usually more actively involved than those in our first cases, wherein the voice recorders featured no or very few contributions from the flight engineers. JAL flight 8054 is an exception, but even there, Yokokawa appeared to communicate unwillingly and without focus. Tragically, in the case of KLM 4805, it was flight engineer Schreuder who saw the dangers of rushing into the takeoff procedure and was aware of the presence of the Pan Am plane. One of the last sentences recorded in the KLM cockpit was his question: "Is he not clear – that Pan American?" But, maybe he assumed that van Zanten knew what he was doing. Moyano, the flight engineer on AVA 052, played the most peculiar role of all in the cases we have looked at. Both passive and aggressive, he seemed to find satisfaction in his own silent recognition of the hopeless situation. Like his colleague Mendenhall, who

---

170 A later reconstruction of the flight revealed that, with the damage it had suffered, the plane was technically no longer controllable.

announced in tones of unmistakable triumph: "that fuel sure went to hell all of a sudden. I told you we had four," Moyano responded to the fuel situation by drawing a finger across his throat and silently indicating the empty fuel gauge to the flight attendant. Obviously, they all failed to make the connection between their observations and their own fates when they were heading for disaster. Randal Thomas is the first integrated and emancipated flight engineer we have encountered. Not only does he participate in the emergency measures, he is also the first person to reassure and encourage Cronin: "You're gonna make this!" His tone is neither importunate nor patronizing; he continues to address Cronin as "captain" but, in this open atmosphere, he feels able to offer a comforting gesture.

The relationship between Cronin and Slader is that of two equals. Their way of speaking to each other is gruff, but it demonstrates concern for the other person. "Make sure we don't hit any fuckin' hills on the way," says Slader to Cronin, and "Put your mask on, Dave." If we analyze their dialogues, we can see that each man is providing the other with a stream of essential information and keeping an eye on what the other is doing. This means they can either confirm the action or correct it if necessary. "Watch your altitude" or "airspeed," warns Slader, and "Watch your heading. You want to go direct to Honolulu."

With this type of interaction, Cronin can express the stress he is feeling and admit he is worried. "I don't know if we're gonna make this. I can't hold altitude." Reassurance is immediately provided by both Thomas and Slader: "We are in good shape. We are in good shape."

The exchange between them on the approach to landing is also a fine example of a truly concerted effort. Slader: "You got the airport?" Cronin: "No." Thomas: "It's right over there to your right." Cronin: "Okay, let's try." There is a further example just before the plane touches down. Cronin to Thomas: "How are you doing?" Thomas: "I'm fine."

As we know, we cannot change the past. However, at this juncture, it is impossible not to wish that the captains in the earlier cases had involved their second officers in this way, or at least bothered to turn round just once and ask, "How are you doing?" Perhaps a gesture of this type, or simply an acknowledgment of his existence, would have been enough for Moyano to say, "Good, but we are going to run out of fuel soon."

The concluding exchange between the cockpit crew of UAL 811 is a fitting finale to their joint effort. Thomas takes charge of the final altitude read-out, Slader adjusts the trim, Cronin gets the plane on the ground, and Slader says, "We're on."[171]

---

171 Kanki, B. and Palmer, M.T. (1993), p. 121.

On May 10, 1989, the crew of UAL flight 811 was presented with the US Department of Transportation's Award for Heroism.[172]

UAL flight 232: get this thing down

Four months later, and before the positive experiences gained from UAL 811 were made public, there was an even more dramatic case involving United Airlines flight 232 (UAL 232). During this flight on July 19, 1989, the main rotor blade of the middle engine of the McDonnell Douglas DC-10 broke.[173] The engine exploded, destroying all of the plane's hydraulic lines. As the DC-10 was designed with three independent and redundant systems, this situation should never have arisen. Although the plane could still fly with the two remaining engines, it could no longer be controlled; without the hydraulic systems neither the aileron, rudder, nor elevator could be moved.[174] Still, the crew managed to land the plane at the airport in Sioux City, Iowa, 44 minutes after the explosion. Controlling this large passenger plane without hydraulic systems was thought to be fundamentally impossible and only succeeded due to some extraordinary teamwork.

The three-strong cockpit crew was already on the third day of a four-day rotation. Captain Alfred "Al" Haynes (57) had been flying for United for 33 years, notching up a total of almost 30,000 flying hours. He had been captaining DC-10s for four years and knew the plane inside out.

With around 20,000 hours, copilot William "Bill" Records (48) was also an extremely experienced pilot. However, he had only been with United for four years and had not yet achieved the rank of captain, despite all his flying hours.[175] Although he had been flying the DC-10 for United for just one year, he had previously flown this type of plane for National Airlines and Pan Am.

---

172 House of Representatives (1989), p. H1798.

173 The following information is based on the report by the National Transportation Safety Board, cf. NTSB (1990), Haynes, A.C (1991), and on an interview with Bill Records.

174 In smaller planes, the control surfaces are operated using mechanical cables. This is not an option in larger planes because of the size of the control surfaces and the force required to manipulate them. The control surfaces on larger planes are therefore operated either hydraulically or electrically.

175 Quite apart from being fundamentally suitable to take command of a plane, the appointment to captain is largely based on seniority, that is, the length of service for that particular airline. As Records only joined United as part of a company takeover, he had low seniority within the company at the time of the accident.

Flight engineer Dudley Dvorak (51) had been working for United for three years. He was also very experienced, having accumulated 15,000 flying hours. However, he had only been flying in the DC-10 for a month.

One of the passengers on board was captain Dennis "Denny" E. Fitch (46), who was flying to Chicago on UAL 232 as an off-duty pilot in the cabin. Like Haynes, he was a veteran and had already been flying for United for 21 years. He had racked up a total flying time of nearly 23,000 hours, of which 2,987 had been spent in the DC-10. What is more, he also worked as a training captain for United in Denver. He was not acquainted with any members of the UAL 232 crew.

The plane was an 18-year-old McDonnell Douglas DC-10–10 that had completed more than 43,000 flying hours and nearly 17,000 rotations. Maintenance work had been carried out as prescribed and no technical defects had occurred.

On July 19, UAL flight 232 left Denver without incident at 2:09 p.m. local time and set course for Chicago (Figure 3.4). On board were 285 passengers and 11 crew members – three in the cockpit and eight in the cabin. It was a clear summer afternoon and Bill Records was the pilot at the

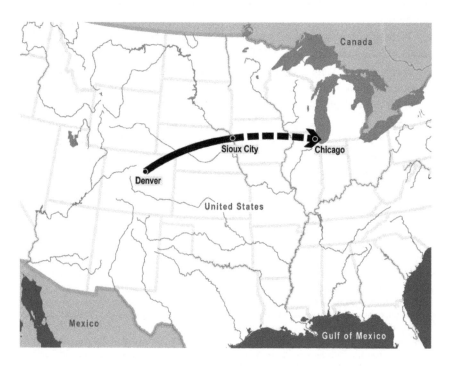

Figure 3.4   **Planned route of UAL 232 from Denver to Chicago**

controls. Captain Haynes was in the middle of drinking a cup of coffee when, at 3:16 p.m.,[176] engine number two (Figure 3.5) exploded.

The pilots in the cockpit heard the explosion, as did the passengers. The plane began to shake violently.[177] Acoustic alerts sounded and warning lights flashed in the cockpit. The plane was shuddering so hard that it was almost impossible to read the instruments, but Haynes and Records finally managed to establish that engine number two had failed. To compensate for this failure, Records applied full thrust to the other engines. The crew pulled out the relevant checklists and worked through them. There are no recordings for the conversations in the cockpit at this time. The voice

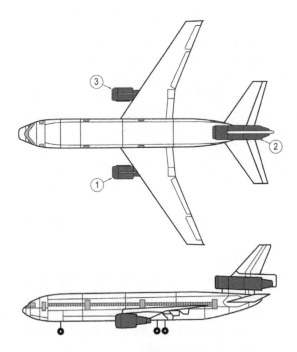

Figure 3.5   **View of the DC-10 showing the engines**

---

176 Central daylight time.

177 The first 10 minutes of this emergency could only be reconstructed afterwards using interviews with the crew. As the voice recorder records information in an endless loop lasting a little more than 30 minutes, only the recordings made after 3:26 p.m. remain.

recorder only records 30 minutes of a flight at a time and is constantly overwritten.

Dvorak realized that the pressure in all three hydraulic systems had fallen to zero. Despite the fully deployed left aileron, the DC-10 kept rolling further and further to the right. If that continued, the plane would end up on its back. As Al Haynes said later, "An airplane about to roll onto its back at 35,000 feet is pretty scary, so you just do anything to make it stop."[178] Haynes reduced the power to engine number one and took control of the plane from Records. The plane returned to the horizontal position.

Not long after, the plane had dropped 2,000 feet (700 m). Although Haynes had managed to stabilize the plane, he knew it was virtually impossible to control it without the hydraulic systems. Records or Dvorak tried to reactivate the hydraulics using a wind-driven auxiliary generator extended below the plane. It did not work, so Haynes had to manipulate the thrust lever extremely carefully to keep the DC-10 even vaguely stabilized.

At 3:20 p.m., Records reported the emergency to Minneapolis Air Route Traffic Control (ARTCC) and requested that the plane be diverted to the nearest airport. The controller at Minneapolis suggested Des Moines International Airport (Iowa), which was 170 miles away. Shortly after, he said that Sioux Gateway Airport would be a better option. Although smaller and with shorter runways, it was only 70 miles away. Air traffic control in Minneapolis handed over to Sioux City. The radar screen in the control room of Sioux City Tower was manned by 27-year-old Kevin Bachman, who had only gained his license for approach control three months earlier.[179]

Minneapolis ARTCC [to Sioux City Approach]: "Sioux City, got an emergency for you."

Sioux City Approach: "All right."

Minneapolis ARTCC: "I've got a United aircraft coming in, lost number two engine, having a hard time controlling the aircraft right now. He is out of twenty-nine thousand (feet) and descending to Sioux City. Right now he is just east of your VOR, but he wants the equipment standing by right now. He's east forty miles."

Sioux City Approach: "Radar contact."

---

178 Haynes, A.C. (1991).
179 Tri-City-Herald (1989), C-3.

Minneapolis ARTCC: "He is having a hard time controlling the plane right now and trying to slow down and get steady on a heading. As soon as I get comfortable, I'll ship him over to you and he'll be in your control."

Sioux City Approach: "All right."

From the cockpit, Dvorak informed the passengers that an engine had failed and the plane would now be heading for an alternate airport. He promised to keep them updated.

Haynes asked purser Janice Brown to come to the cockpit and explained the situation to her. He told her she would have to prepare the passengers for an emergency landing. Brown returned to the cabin and passed the information on to the other flight attendants. Next, training captain Denny Fitch, who was in the first-class section, spoke to Brown and offered to help out in the cockpit. Brown promptly got back to Haynes and asked whether that would be okay. Haynes said, "Bring him up." By this time, he and Records barely managed to keep the plane leveled. At 3:26 p.m., Haynes reported to Bachman in Sioux City. There are recordings of the conversations in the cockpit from this point on. The following parts are extracts, minus an overload of technical details.

Haynes [to Sioux City Approach]: "Sioux City approach, United two thirty two heavy, with you out of twenty-six [thousand feet]. We got about a five-hundred-foot rate of descent. Okay, you know, we have almost no control ability, very little elevator, almost no aileron. We're controlling the turns by power. I don't think we can turn right. I think we can only make left turns."

Sioux City Approach: "United two thirty two heavy, understand you can only make right turns."[180]

Haynes: "That's affirmative."

Haynes [to Records]: "Now that goddamn elevator doesn't want to work. Rolling right."

Records: "Very little elevator. It's hard or sluggish. Al, do you want me to slew this elevator?"

Haynes: "Yeah, do whatever you can."

---

180 Bachman already saw on his radar screen that UAL 232 was only turning right. He just asked Haynes for a confirmation and therefore clarified Haynes' slightly confused radio transmission.

Despite both their efforts, the DC-10 continued to rise and drop in a two-minute cycle.[181] Each time the plane dropped a further 2,000 feet in altitude. For the passengers, it was like riding a rollercoaster.

By this time, Dvorak had established radio contact with the United maintenance center in San Francisco (San Francisco Aero Maintenance, SAM). He wanted to know how they could stabilize the plane. Perhaps he was too matter-of-fact in describing their dilemma because the engineers from SAM did not appear to grasp the full severity of the situation. Also, for such situations there were no emergency procedures in place; the loss of hydraulic power was thought impossible due to a triple redundant system. Should such a loss occur, it was considered hopeless for all involved. Al Haynes later described this failure as a further example of "Murphy's Law."[182] No emergency procedures were in place for such situations.

> Dvorak [to SAM]: "This is United two thirty two heavy. We blew number two engine and we've lost all hydraulics and we are only able to control level flight with the asymmetrical power settings. We have very little rudder or elevator."
>
> SAM: "United two thirty two, understand you lost number two engine totally, sir."
>
> Dvorak: "That's affirmative."
>
> SAM: "System one and system three? They are operating normally?"
>
> Dvorak: "Negative. All hydraulics are lost."
>
> SAM: "Okay, United two thirty two, where you gonna set down?"
>
> Dvorak: "We need some assistance right now. We can't – we are having a hard time controlling it."
>
> SAM: "Okay, United two thirty two. I'll try to help you. I'll pull out your flight manual."
>
> Records: "Wonder about the outboard ailerons. If we put some flaps out, you think that would give us outboard?"
>
> Dvorak: "God, I hate to do anything."
>
> Haynes: "Well, we're going to have to do something."

---

181 Known as phugoid oscillations, these occurred due to incorrect trimming following the engine failure and could no longer be corrected due to the failure of the hydraulic systems. The crew managed to limit these oscillations by adjusting the engine power.

182 Haynes, A.C. (1991).

The time was now 3:29 p.m. It had been 13 minutes since the explosion. In the tower of Sioux City, Bachman had ordered preparations for an emergency landing. On the radar screen he could see that the plane was now nearly impossible to control. The DC-10 kept wallowing to the right. Bachman spoke to his supervisor, Mark Zielezinski, and told him they would probably lose the plane. In San Francisco, the engineers at SAM tried to help the crew of UAL 232. The crisis team at United had also been notified. A rescue team was sent to Sioux City. The fire service at the airport had put the hospitals in Sioux City on alert.

Fitch entered the cockpit. Without pausing to greet him, Haynes asked Fitch to see what he could see from the cabin. Fitch returned to the cabin.

Records: "Don't pull the throttles off. What's the hydraulic quantity?"

Dvorak: "Down to zero."

Records: "On all of them?"

Dvorak: "All of them."

Haynes: "Quantity is gone?"

Dvorak: "Yeah, all the quantity is gone. All pressure is gone."

Haynes: "You got hold of SAM?"

Dvorak: "Yeah, I've talked to him."

Haynes: "What's he saying?"

Dvorak: "He's not telling me anything."

Fitch now reentered the cockpit. Haynes turned to him.

Haynes: "We've lost – no hydraulics. We have no hydraulic fluid. That's part of our main problem."

Fitch: "Okay. Both your inboard ailerons are sticking up. That's as far as I can tell."

Haynes: "Well that's because we're steering – we are turning – maximum turn right now."

Fitch: "Tell me. Tell me what you want and I'll help you."

Haynes: "Right throttle. Close one, put two up. What we need is elevator control. And I don't know how to get it."

Fitch: "Okay."

Haynes asked Fitch to help them control the remaining two engines. Fitch took up position at the middle console between Haynes and Records. Haynes focused on coordinating other activities. Dvorak received a radio communication from United Dispatch, which had just been informed of the emergency.

> Dispatch: "United two thirty two, do you want to put that thing on the ground right now or do you want to come to Chicago?"
>
> Dvorak: "Well, we can't make Chicago. We're gonna have to land somewhere out here, probably in a field."

Haynes turned to Fitch. "How are they doing on the evacuation?"

> Fitch: "They're putting things away, but they're not in a big hurry."
>
> Haynes: "They better hurry. We're gonna have to ditch, I think."
>
> Fitch: "Yeah."
>
> Haynes: "Okay, I don't think we're going to make the airport."
>
> Records: "No. We got no hydraulics at all."
>
> Haynes: "Gotta put some flaps and see if that'll help."
>
> Records: "You want them now?"
>
> Haynes: "What the hell. Let's do it. We can't get any worse than we are."
>
> Records: "Slats are out."
>
> Fitch: "No, you don't have any slats."
>
> Haynes: "We don't have any hydraulics, so we're not going to get anything."
>
> Sioux City Approach: "United two thirty two heavy, can you hold that present heading, sir?"
>
> Fitch: "Ask them where the hell we are."
>
> Haynes [to Sioux City Approach]: "Where's the airport to us now, as we come spinning down here?"
>
> Sioux City Approach: "United two thirty two heavy, Sioux City airport is about twelve o'clock and three six miles."
>
> Haynes: "Okay we're trying to go straight. We're not having much luck."

Fitch: "Okay, if you get denser air, you should get level flight back again. Whatever you got you got."

Haynes [laughing]: "We didn't do this thing on my last performance check in a simulator."

All four laugh.

Haynes: "I poured coffee all over – it's just coffee. We'll get this thing on the ground. Don't worry about it."

Records: "It seems controllable, doesn't it, Al?"

Fitch: "Yeah. The lower you get, the more dense the air is and the better your shots. Okay?"

Haynes [to Sioux City Approach]: "Sioux City, United two thirty two, could you give us please your ILS frequency, the heading, and length of the runway?"

Sioux City Approach: "Two thirty two heavy, the localizer frequency is one zero nine point three. It'll take about two four zero heading to join it. The runway is nine-thousand-feet long."

Thanks to the lower altitude and therefore denser air, Fitch, Haynes, and Records had now managed to stabilize the DC-10, although the intermittent rise and fall continued.

SAM: "United two thirty two, this is SAM."

Dvorak: "SAM, two thirty two, we're gonna try and put into Sioux City. We're very busy right now. We're trying to go into Sioux City. We'll call you as soon as I can."

SAM [to Dispatch]: "He has no control. He's using that kind of sink rate, I believe. This is what he's doing. He's got his hands full for sure."

Haynes: "Start dumping fuel. Just hit the quick dump. Let's get the weight down as low as we can."

Dvorak: "I didn't have time to think about that."

Haynes: "Try not to lose any more altitude than we have to."

Records: "Okay. Go ahead and dump."

Fitch: "This thing seems to want to go right more than it wants to go left, doesn't it?"

Sioux City Approach: "United two thirty two heavy, did you get the souls on board count?"

Haynes [to Dvorak]: "What did you have for a count for people?"

Haynes [to Sioux City Approach]: "Let me tell you, right now we don't even have time to call the gal."

Dvorak [to Sioux City Approach]: "Two ninety two."[183]

Fitch: "Power's coming back in."

Haynes: "Bring it to the right one. You got to go left. We just keep turning right, still turning right."

Fitch: "That's what I am trying to do."

Haynes [to Sioux City Approach]: "Two thirty two, we're just gonna have to keep turning right. There's not much we can do about turning left. We'll try to come back around to the heading."

Records: "Is that Sioux City down to the right?"

Haynes: "That's Sioux City."

Haynes [to Dvorak]: "Did you ever get hold of SAM?"

Dvorak: "Yep. Didn't get any help."

Records [to Sioux City Approach]: "Where is Sioux City from our present position, United two thirty two?"

Sioux City Approach: "United two thirty two, it's about twenty on the heading and thirty-seven miles."

Haynes to Fitch: "You had the thing leveled off for minute. My name's Al Haynes."

Fitch: "Hi, Al. Denny Fitch."

Haynes: "How do you do, Denny?"

Fitch: "I'll tell you what. We'll have a beer when this is all done."

Haynes: "Well, I don't drink, but I'll sure have one."

Fitch: "You lost the engine?"

Haynes: "Yeah, well, yeah. It blew. We couldn't do anything about it. We shut it down."

Fitch: "Yeah."

---

183 There were in fact 296 people on board – 285 passengers and 11 crew members.

Haynes: "Can't think of anything that we haven't done. There really isn't a procedure for this."

SAM: "United two thirty two, in your handbook page ninety-one."

Dvorak: "We already have a no flap, no slat made up and we're getting ready. We're gonna try to put into Sioux City with gear down."

Sioux City Approach: "When you get turned to that two-forty heading, sir, the airport will be about twelve o'clock and thirty-eight miles."

Records: "Okay, we're trying to control it just by power alone now. We have no hydraulics at all, so we're doing our best here."

Sioux City Approach: "Roger, and we've notified the equipment out in that area, sir. The equipment is standing by."

Haynes called Janice Brown into the cockpit.

Haynes: "Everybody ready? We've almost no control of the airplane. It's gonna be tough, gonna be rough."

Brown: "So we're gonna evacuate?"

Haynes: "Yeah. And if we can keep the airplane on the ground and stop standing up, give us a second or two before you evacuate. 'Brace' will be the signal; it'll be over the PA system – 'brace, brace, brace.'"

Brown: "And that will be to evacuate?"

Haynes: "No, that'll be to brace for landing. And if you have to evacuate, you'll get the command signal to evacuate, but I really have my doubts you'll see us standing up, honey. Good luck, sweetheart."

Brown: "Thank you." The purser returned to the cabin.

Dvorak: "She [Janice Brown] says there appears to be some damage on that one wing. Do you want me to go back and take a look?"

Fitch: "No, we don't have time."

Haynes: "Okay, go ahead. Go ahead and see what you can see, not that it'll do any good. I wish we had a little better control of the elevator. They told us the autopilot would do this, but it sure as hell won't. Try yours again."

Fitch: "All right, we came into the clean air."

Haynes: "Turn, baby."

Fitch: "Which way do you want it, Al?"

Haynes and Records: "Left."

Haynes: "Back on that sucker down a little bit more."

Haynes: "How do we get the gear down?"

Fitch: "Well, they can freefall. We got the gear doors down?"

Haynes: "Yep."

Records: "We're gonna have trouble stopping too."

Haynes: "Oh yeah. We don't have any brakes."

Fitch: "Braking will be a one-shot deal. Just mash it, mash it once. That's all you get. I'm gonna give you a left turn back to the airport. Is that okay?" [184]

Haynes [to Sioux City]: "Okay, United two thirty two, we're starting to turn back to the airport. Since we have no hydraulics, braking is gonna really be a problem. I would suggest the equipment be toward the far end of the runway. I think under the circumstances, regardless of the condition of the airplane when we stop, we're going to evacuate. So you might notify the ground crew that we're gonna do that."

Sioux City Approach: "United two thirty two heavy, wilco,[185] sir. If you can continue that left turn to about two-twenty heading, sir, that'll take you to the airport."

Haynes: "Two-twenty, roger."

Haynes [to Dvorak]: "What did SAM say? Good luck?"

Dvorak: "He hasn't said anything."

Haynes: "Okay, we'll forget them. Tell them you're leaving the air and you're gonna come back up here and help us...and screw them. Ease her down just a little bit."

At 3:49 p.m., Haynes gave the order to deploy the landing gear.

Haynes: "Okay, put it [the gear] down."

Fitch: "Got to get my glasses on or can't see shit."

Records [to Sioux City Approach]: "Where's the airport?"

Sioux City Approach: "United two thirty two, the airport's currently twelve o'clock and two one miles. You're gonna have to widen out just

---

184 This is odd given that the crew was convinced they could not make any left turns. After the flight, none of the four crew members could remember having flown a left turn. However, both the voice recorder and radar recordings provided evidence of this.

185 Used in radio communications to confirm "will comply."

a little to your left, sir, to make the turn to final and also it'll take you away from the city."

Haynes [to Sioux City Approach]: "Whatever you do, keep us away from the city."

Sioux City Approach: "You are currently one seven miles north-east of the airport. You're doing good."

Dvorak was having trouble coping with controlling the engines and turned to Fitch.

Dvorak: "Do you want this seat?"

Fitch: "Yes, do you mind?"

Dvorak: "I don't mind. I think you know what you're doing there."

Fitch and Dvorak swapped places.

Haynes: "Want to keep turning right. Want to go to the airport."

Fitch: "I got the tower."

Haynes: "Come back, all the way back."

Fitch: "I can't handle that steep a bank."

Haynes: "Damn it. Wish we hadn't put that gear down."

Dvorak: "Ah, well."

Fitch: "We don't know."

Haynes: "I want to get as close to the airport as we can."

Fitch: "Okay."

Haynes: "If we have to set this thing down in dirt, we set in the dirt."

Dvorak [over PA]: "We have four minutes to touchdown, four minutes to touchdown."

Sioux City Approach: "United two thirty two heavy, roger. Can you pick up a road or something up there?"

Fitch: "Airport's down there. Got it."

Haynes: "I don't see it yet."

Fitch: "No left at all."

Haynes: "Back, back – forward, forward. Won't this be a fun landing?" [Laughter]

Sioux City Approach: "United two thirty two heavy, roger. The airport's currently at your one o'clock position, one zero miles."

Fitch: "I got the runway."

Haynes: "I don't. Come back, come back."

Fitch: "It's off to the right over there."

Sioux City Approach: "United two thirty two heavy, if you can't make the airport, sir, there is an interstate that runs north to south, to the east side of the airport. It's a four-lane interstate."

Haynes [to Sioux City Approach]: "We have the runway in sight. We'll be with you shortly. Thanks for your help."

Fitch: "Bring it down. Ease her down."

Records: "Oh, baby."

Sioux City Approach: "United two thirty two heavy, the wind's currently three six zero [degrees] at eleven [knots]. You are cleared to land on any runway."

Haynes [laughing]: "Roger. You want to be particular and make it a runway, huh?"

Sioux City Approach: "There's a runway that's closed, sir. That could probably work."

Haynes [to Sioux City Approach]: "We're pretty well lined up on this one here."

Two minutes before the landing, Bachman realized that UAL 232 was heading for runway 22 rather than the planned runway 31 that had been cleared in preparation. As it was not in use, runway 22 was currently occupied by all the fire vehicles and heavy equipment waiting for UAL 232 to land on runway 31. Bachman immediately redirected the rescue teams and cleared runway 22. Haynes and Records noticed the vehicles scrambling to the side.

Sioux City Approach: "United two thirty two heavy, roger, sir. That's a closed runway, sir, that'll work, sir. We're getting the equipment off the runway."

Haynes [to Sioux City Approach]: "How long is it?"

Sioux City Approach: "Sixty-six-hundred feet, six thousand six hundred. Equipment's coming off."

Haynes [to Fitch]: "Pull the power back. That's right – pull the left one back."

Records: "Pull the left one back."

Sioux City Approach: "At the end of the runway it's just wide-open field."

Dvorak [on PA]: "Brace, brace, brace."

Haynes: "Close the throttles."

Records: "Left Al. Left, left, left – we're turning."

The impact is audible at 4:00 p.m. The voice recorder stopped.

The plane hit the runway with its right wing and right landing gear. The tail sheared off, the plane slid along the runway and broke into several pieces. The fuel remaining in the wings ignited, creating a massive fireball. A local TV station that had happened to hear of the situation had sent a camera team to the airport. They filmed the plane as it approached and then broke apart on the runway: 111 passengers and one flight attendant died; 184 people – most of them uninjured or only lightly injured – survived (Figures 3.6 and 3.7).

Figure 3.6 **Flight path of UAL 232**

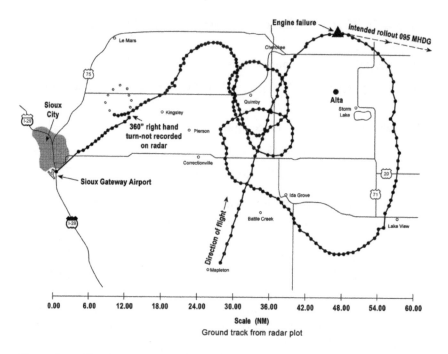

Figure 3.7   **Wreckage path of UAL 232**

Almost 45 minutes had passed before the firefighters became aware of the pilots stuck in the demolished front section of the aircraft. Al Haynes later described what happened in the cockpit: "[...] the four of us are right there. That area is normally about ten feet high. [...] In the rescue operation, they came in and tried the jaws of life. They put it up on Bill's side, and as they did it put pressure on my side. And I happened to be conscious at the time. And I strongly recommended they stop doing it. So they came to my side, and tried my side, and Bill did the same thing. Bill's seat had collapsed, the back of the seat, with him inside it. He had eight broken ribs [...] and a broken pelvis. So he was in a little bit of pain. So what they did was they came up with the idea to bring a fork-lift over, and run the chains [...] and lift the cockpit straight up. And by doing this raised the cockpit to this height, and pulled us all out of the bottom. That's how they got us out of the airplane (Figure 3.8)."[186]

The NTSB investigation into the accident concluded that the main rotor disc of engine two had broken due to material fatigue. Pieces of

---

186 Haynes, A.C. (1991).

Figure 3.8   **Al Haynes at the first press conference after the crash**

debris ejected at high speed destroyed all three hydraulic lines that, in this particular part of the plane, lie together largely unprotected. The NTSB later drew up proposals for ways to improve protection for the planes' control systems,[187] and these were subsequently implemented by McDonnell Douglas. In addition, McDonnell Douglas' test pilots developed a series of recommendations for how to control DC-10 planes in similar emergency situations.[188]

The NTSB reserved particular praise for the crew of UAL flight 232. Based on its analysis, it should have been impossible for them to make the emergency landing in Sioux City. The CRM crisis management displayed by the UAL 232 crew is still viewed today as exemplary. In a speech given at the NASA Ames Research Center,[189] Al Haynes summed up the reasons for the successful emergency landing as follows: "I think there are five factors that contributed to the degree of success that we had at Sioux City: that is, luck, communications, preparation, execution, and cooperation."

---

187 NTSB (1990), pp. 102–109.
188 Ibid., pp. 116–123.
189 Haynes, A.C. (1991).

One of the factors that made Haynes and the other survivors so *lucky* was the fact that they were flying over Iowa and not Manhattan or the Pacific. "We knew in the back of our mind that where we were, if we had to ditch, we could probably find some fairly flat land, and we might have a chance of survival."[190] The weather also played a decisive role, as did the time of day. On that clear afternoon in July 1989, the crew was able to spot the airport at Sioux City from a distance of five miles.

In terms of *communication*, Haynes singled out his interaction with Bachman, "this extremely calm young man," who was so focused in providing support, information, and suggestions. In hindsight, Al Haynes even cut his helpless colleagues at SAM some slack. "When I first had Dudley stop communicating with them and turn around for the landing, I was a little ticked, until I realized how frustrating it must have been for these four or five people, there, with all those computers, with all the knowledge at their fingertips [...] and there's absolutely nothing they could do to help a crew. [...] I have not yet had a chance to go down and see them and apologize for what I was thinking, but at least I didn't say it out loud."[191]

In his speech at NASA, Haynes also addressed the external communications that took place between Bachman and the teams at the airport and in the nearby hospitals, which I have not documented here, as they exceed our area of interest. The same applies to Haynes' reference to cooperation, meaning the concerted efforts of the external support teams. Let us just think for a moment what it takes to spring into action upon receiving an Alert 3 saying "an airplane has crashed," alert fire service, rescue teams, and post-traumatic stress units in order to instigate a rescue mission, care for the survivors, and treat the next of kin of those who died with due sympathy and respect.

With regard to *preparation*, as Haynes rightly says, there was none: "But the preparation that paid off for the crew was something that United started in 1980 called Cockpit Resource Management. [...] Up until 1980, we kind of worked on the concept that the captain was THE authority on the aircraft. What he says, goes. And we lost a few airplanes because of that. And we would listen to him and do what he said and we wouldn't know what he's talking about. [...] So if I hadn't used CRM, if we had not let everybody put their input in, it's a clinch we wouldn't have made it. [...] The days of the captain being the ultimate authority are gone. He may be

---

190 Ibid.
191 Ibid.

the authority on the airplane, he may sign for the papers and all this, but you don't work that way."[192]

Haynes' description of *execution* has an almost literary quality and illustrates the interaction between Haynes, Records, and Dvorak even more clearly than the voice recordings:

"We've lost several airplanes because everybody was working on the problem and nobody was flying the airplane. [...] But somebody has got to fly the airplane. Bill immediately took hold of the airplane, immediately called ATC and said we lost an engine and had to get a lower altitude [...] all those things you're supposed to do. So my attention now is diverted do Dudley [...]. Dudley got out his book and the first thing it said was, close the throttle. And when I tried to pull the throttle back, it wouldn't come back. Now, I've never shut an engine down in flight on a jet, so I didn't know that when you pulled the throttle back, it didn't come back. In the simulator when you do it, it always came back. [...] Dudley says, well, try the fuel. [...] I tried to shut the fuel off, and the fuel lever wouldn't move. [...] We did get the fuel shutoff by pulling down the firewall shutoff; which shut off all the electrics and hydraulics to the engine. And then the fuel went off, whether it was coincidental, or I had actually helped it, I don't know. [...] And Bill says to me, Al, I can't control the airplane. [...] This is when I said the dumbest thing I've ever said in my life, I said 'I got it.' I didn't have it very long."[193]

As it happens, years later, the cockpit crew of the DHL Airbus OO-DLL took a leaf out of the UAL 232 crew's book when their plane was hit by a ground-to-air missile fired by Iraqi insurgents following takeoff from Baghdad on November 22, 2003.[194] In this case, the left wing was badly damaged and caught fire, and the three hydraulic lines were also severed. Like Al Haynes, captain Eric Gennotte and his crew managed to fly back and make an emergency landing in Baghdad using just the engine throttle power. Among other things, he and his crew, too, attributed their ability to manage the crisis to their CRM training.[195]

I want to come back to the increased performance capacity of CRM teams and stress that the disasters we dealt with in the first part would have been avoidable had the cockpits been fear-free environments that promoted cooperation. In contrast, in the cockpits of UAL flights 811 and 232 – where everyone had a reason to panic – the psychological

192 Ibid.
193 Ibid.
194 Rosay, J. (2004).
195 Lutz, T. (2004), p. 33.

stability of those in charge made it possible to rescue a situation that could have ended fatally for everyone on board. Nearly everyone on UAL 811 survived, as did more than 60 percent of those on board UAL 232. Naturally, not every aspect of the achievements of the two crews can be ascribed to CRM. Al Haynes mentioned the other factors in his speech at NASA. Nonetheless, CRM training facilitated the flow of information in the cockpit. As to UAL 232, during the most intense moments, there were 60 notifications and comments per minute.[196] I have listed one after another, but in reality they were constantly overlapping. Their basic purpose was to communicate information, but they were also a valuable way to let off steam ("Turn, baby. Back on that sucker down a little bit more") or to express annoyance ("Damn it. Wish we hadn't put that gear down") or provide encouragement ("You're gonna make this"). Sometimes they provide corrections ("Don't get any lower, captain"), or ("Pull the power back. That's right – pull the left on back"). Or they lighten the mood ("Back, back – forward, forward. Won't this be a fun landing?") as well as offer a way to provide reassurance ("We'll get this thing on the ground. Don't worry about it"). They help with expressing fear ("God, I hate to do anything"). They are used to ask for help ("Al, I can't control the airplane"), to regroup seconds before the landing ("Bring it down. Ease her down"), and as a way to seek comfort ("It seems controllable, doesn't it, Al?").

The humor employed by captains Cronin and Haynes is not reproducible, neither is it an essential leadership quality, even if it can smooth the path of cooperation and interaction. However, what is clearly evident is their air of egalitarianism and the way they create an environment where communication is open and free from any fear or inhibition – instead of insisting on ruling the roost.[197]

CRM reporting system

As we have already seen, implementing the CRM concept was tough to begin with. Although there were pilots who accepted it after a short time, it took the majority years to adjust. That all changed after the publication of the investigation reports on UAL flights 811 and 232. In these cases, two planes and more than 400 passengers had been saved in situations that

---

196 Helmreich, R.L. (1994), p. 275.
197 The communication of the UAL crew has been described as a prime example for swift
    action teams in managing an extreme crisis situation. McKinney, E.H. et al. (2005).

had looked like inevitable disasters, and the respective captains attributed part of their achievement to CRM. It is quite possible that it was their statements rather than the training sessions and courses that ultimately made CRM acceptable to pilots. Whatever the reason, at the beginning of the 1990s, more than 90 percent of pilots stated that they found CRM training extremely helpful or very helpful.[198] The number of those who rejected it continued to fall.

The success of CRM is also evident from the evaluation of accident reports. A study by Baker, Qiang, Rebok, and Li[199] analyzed a total of 558 NTSB accident reports in the United States between 1983 and 2002, and revealed that the number of pilot errors contributing toward accidents dropped by 25 percent in the period from 1983 to 1987, that is, following the introduction of CRM. By 2002, it had fallen by 40 percent.

One problem remained: CRM effectiveness could still only be assessed by using questionnaires and counting the decreasing number of air accidents. The cases we have examined featured reports produced by the investigation commissions of the relevant national aviation authorities, mainly the NTSB in the United States. They all contain the detailed causes for accidents and crew evaluation analyses. However, these are always based on emergency situations; interpretations of final conversation snippets caught on the voice recorder; statements from surviving crew members, passengers, air traffic controllers, and other witnesses, if available, and independent expert analyses. These sources were not enough to provide a comprehensive evaluation of CRM or to develop it further. As a result, aviation authorities and airlines needed an additional means of gathering information about those errors that had not actually caused an accident but could have.

It was probably an obvious step to set up a system that would enable pilots, flight attendants, air traffic controllers, and maintenance staff to report errors and their causes. As incorrect actions were almost always at the root of the problem, those in charge realized that this reporting system had to exclude the possibility of people denouncing each other and ensure that no sanctions would be imposed on those involved. Hence the rule that the person making the report supplies his or her own name, but not the name of the person who is the subject of the report (Figure 3.9). As we will see, even airport names and flight numbers remain anonymous or are excluded.

---

198 Helmreich, R.L. and Foushee, H.C. (1993), p. 33ff.
199 Baker, S.P. et al. (2008).

B

DO NOT REPORT AIRCRAFT ACCIDENTS AND CRIMINAL ACTIVITIES ON THIS FORM.
ACCIDENTS AND CRIMINAL ACTIVITIES ARE NOT INCLUDED IN THE ASRS PROGRAM AND SHOULD NOT BE SUBMITTED TO NASA.
ALL IDENTITIES CONTAINED IN THIS REPORT WILL BE REMOVED TO ASSURE COMPLETE REPORTER ANONYMITY.

(SPACE BELOW RESERVED FOR ASRS DATE/TIME STAMP)

IDENTIFICATION STRIP: *Please fill in all blanks to ensure return of strip.*
NO RECORD WILL BE KEPT OF YOUR IDENTITY. This section will be returned to you.

TELEPHONE NUMBERS where we may reach you for further
details of this occurrence:

HOME   Area _____ No. _____   Hours _____

WORK   Area _____ No. _____   Hours _____

TYPE OF EVENT/SITUATION

NAME _____

ADDRESS/PO BOX _____

_____

DATE OF OCCURRENCE _____
                         (MM/DD/YYYY)

CITY _____ STATE _____ ZIP _____   **LOCAL TIME (24 hr. clock)** _____
                                                 (HH:MM)

PLEASE FILL IN APPROPRIATE SPACES AND CHECK ALL ITEMS WHICH APPLY TO THIS EVENT OR SITUATION.

REPORTER		FLYING TIME (in hours)	CERTIFICATES & RATINGS		ATC EXPERIENCE	
☐ Captain	☐ Single Pilot	Total Time _____ hrs	☐ Student	☐ Flight Instructor	☐ FPL	☐ Developmental
☐ First Officer	☐ Instructor		☐ Sport/Rec	☐ Multiengine	radar _____ yrs	
☐ pilot flying	☐ Trainee	Last 90 Days _____ hrs	☐ Private	☐ Instrument	non-radar _____ yrs	
☐ pilot not flying	☐ Dispatcher: _____ yrs		☐ Commercial	☐ Flight Engineer	supervisory _____ yrs	
☐ relief pilot	☐ Other: _____	Time in Type _____ hrs	☐ ATP	☐ Other: _____	military _____ yrs	
☐ check airman						

AIRSPACE		CONDITIONS/WEATHER ELEMENTS			LIGHT/VISIBILITY		ATC/ADVISORY SVC.	
☐ Class A	☐ Class E	☐ VMC	☐ fog	☐ snow	☐ dawn	☐ night	☐ Ramp	☐ Center
☐ Class B	☐ Class G	☐ IMC	☐ hail	☐ thunderstorm	☐ daylight	☐ dusk	☐ Ground	☐ FSS
☐ Class C	☐ Special Use	☐ Mixed	☐ haze/smoke	☐ turbulence	Ceiling _____ feet		☐ Tower	☐ UNICOM
☐ Class D	☐ TFR	☐ Marginal	☐ icing	☐ windshear	Visibility _____ miles		☐ TRACON	☐ CTAF
			☐ rain	☐ other: _____	RVR _____ feet		ATC Facility Name:	

AIRCRAFT 1				AIRCRAFT 2		
Your Aircraft Type (Make/Model) (e.g. B737) NOT "N #", FR #, etc.: _____		Operating FAR Part: _____	Other Aircraft: _____			Operating FAR Part: _____
Operator	☐ air carrier ☐ air taxi ☐ corporate	☐ fractional ☐ FBO ☐ government	☐ military ☐ personal ☐ other: _____	☐ air carrier ☐ air taxi ☐ corporate	☐ fractional ☐ FBO ☐ government	☐ military ☐ personal ☐ other: _____
Mission	☐ passenger ☐ personal	☐ cargo/freight ☐ training	☐ ferry ☐ other: _____	☐ passenger ☐ personal	☐ cargo/freight ☐ training	☐ ferry ☐ other: _____
Flight Plan	☐ VFR ☐ IFR	☐ SVFR ☐ DVFR	☐ none	☐ VFR ☐ IFR	☐ SVFR ☐ DVFR	☐ none
Flight Phase	☐ taxi ☐ parked ☐ takeoff ☐ initial climb	☐ climb ☐ cruise ☐ descent ☐ initial approach	☐ final approach ☐ missed/GAR ☐ landing ☐ other: _____	☐ taxi ☐ parked ☐ takeoff ☐ initial climb	☐ climb ☐ cruise ☐ descent ☐ initial approach	☐ final approach ☐ missed/GAR ☐ landing ☐ other: _____
Route in Use	☐ airway (ID): _____ ☐ direct ☐ SID (ID): _____	☐ STAR (ID): _____ ☐ oceanic ☐ vectors	☐ visual approach ☐ none ☐ other: _____	☐ airway (ID): _____ ☐ direct ☐ SID (ID): _____	☐ STAR (ID): _____ ☐ oceanic ☐ vectors	☐ visual approach ☐ none ☐ other: _____

If more than two aircraft were involved, please describe the additional aircraft in the "Describe Event/Situation" section.

LOCATION	CONFLICTS	
Altitude: _____ (single value) ☐ MSL ☐ AGL	Estimated miss distance in feet: horiz _____ vert _____	
Distance: _____ and/or Radial (bearing): _____ from: _____	Was evasive action taken?	◉ Yes ◉ No
☐ Airport _____ ☐ ATC Fac _____	Was TCAS a factor?	◉ TA ◉ RA ◉ No
☐ Intersection _____ ☐ NAVAID _____	Did terrain warning system activate? [Reset]	◉ Yes ◉ No

NASA ARC 277B (May 2009)   GENERAL FOR   Page 1 of 3

Figure 3.9a   **ASRS report sheet**

NATIONAL AERONAUTICS AND SPACE ADMINISTRATION

NASA has established an Aviation Safety Reporting System (ASRS) to identify issues in the aviation system which need to be addressed. The program of which this system is a part is described in detail in FAA Advisory Circular 00-46D and FAA Handbook 7210.3. Your assistance in informing us about such issues is essential to the success of the program. Please fill out this form as completely as possible, enclose in a sealed envelope, affix proper postage, and send it directly to us.

The information you provide on the identity strip will be used only if NASA determines that it is necessary to contact you for further information. THIS IDENTITY STRIP WILL BE RETURNED DIRECTLY TO YOU. The return of the identity strip assures your anonymity.

AVIATION SAFETY REPORTING SYSTEM

Section 91.25 of the Federal Aviation Regulations (14 CFR 91.25) prohibits reports filed with NASA from being used for FAA enforcement purposes. This report will not be made available to the FAA for civil penalty or certificate actions for violations of the Federal Air Regulations. Your identity strip, stamped by NASA, is proof that you have submitted a report to the Aviation Safety Reporting System. We can only return the strip to you, however, if you have provided a mailing address. Equally important, we can often obtain additional useful information if our safety analysts can talk with you directly by telephone. For this reason, we have requested telephone numbers where we may reach you.

Thank you for your contribution to aviation safety.

NOTE: AIRCRAFT ACCIDENTS SHOULD NOT BE REPORTED ON THIS FORM. SUCH EVENTS SHOULD BE FILED WITH THE NATIONAL TRANSPORTATION SAFETY BOARD AS REQUIRED BY NTSB Regulation 830.5 (49CFR830.5).

If you want to mail this form, please fold both pages (and additional pages if required), enclose in a sealed, stamped envelope, and mail to:

NASA AVIATION SAFETY REPORTING SYSTEM
POST OFFICE BOX 189
MOFFETT FIELD, CALIFORNIA 94035-0189

DESCRIBE EVENT/SITUATION

Keeping in mind the topics shown below, discuss those which you feel are relevant and anything else you think is important. Include what you believe really caused the problem, and what can be done to prevent a recurrence, or correct the situation. (USE ADDITIONAL PAPER IF NEEDED)

CHAIN OF EVENTS	Page 2 of 3	HUMAN PERFORMANCE CONSIDERATIONS
· How the problem arose · How it was discovered		· Perceptions, judgments, decisions · Actions or inactions
· Contributing factors · Corrective actions		· Factors affecting the quality of human performance

NASA ARC 277A (May 2009)

Figure 3.9b **ASRS report sheet**

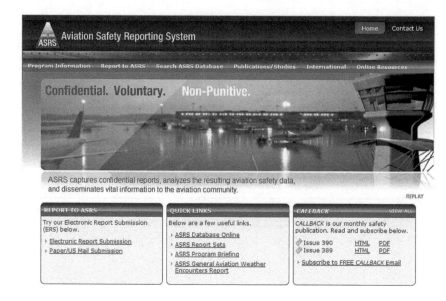

Figure 3.10 **Homepage of NASA's ASRS**

NASA was the driving force in setting up the Aviation Safety Reporting System (ASRS). It would become the official reporting system as early as 1976 and openly guaranteed "confidential, voluntary, non-punitive" reporting for everyone (Figure 3.10). This freedom from sanctions is furthermore determined by law.[200] Even in cases where regulations have been breached, disciplinary proceedings can be averted if the breach is reported voluntarily.[201] The only time this exemption from prosecution does not apply is in cases of extremely serious and deliberate violations, such as flying under the influence of alcohol or drugs, but even in these cases NASA will not forward any information to the FAA.

The ASRS reporting system was later adopted by aviation authorities in other countries. It was introduced in the United Kingdom in 1982, for example, where it goes by the name Confidential Human Factors Incident Reporting Programme (CHIRP) (Figure 3.11).

---

200 Title 14 of the Code of Federal Regulations (14 CFR) part 91, § 91.25. Since the beginning of the program, no personal data has been passed on to the FAA.
201 Title 49 of the United States Code (49 U.S.C.) subtitle VII.

Figure 3.11   **Reporting guidelines CHIRP**

The International Confidential Aviation Safety Systems (ICASS) Group is responsible for coordinating the national reporting systems. This has generated a number of comprehensive databases. Additional reporting systems have also been established between specific aviation authorities and airlines, such as the FAA's Aviation Safety Action Program (ASAP) in the United States.[202] The reports cover everything from oversights and mistakes to near misses that would never have been recorded in regular analysis reports (Figure 3.12).

---

202 All the major US airlines participate in this program. As with the ASRS program, pro-
    viding evidence of ASAP reports being submitted can avert disciplinary measures by the
    FAA in situations where regulations have been breached.

ASRS June 2012 Report Intake	
Air Carrier/Air Taxi Pilots	3,807
General Aviation Pilots	1,181
Controllers	856
Cabin	282
Mechanics	135
Dispatcher	117
Military/Other	24
TOTAL	6,402

Figure 3.12   **Monthly ASRS report intake, June 2012**
Note: On average, around 5,000 reports are received each month.

Airlines apply the information received to devise training scenarios for use in simulators and workshops. These scenarios are to be analyzed and implemented to correct or prevent erroneous or detrimental behavior patterns. Aircraft manufacturers in turn use the information to further develop their planes and instruments. In addition, NASA sorts and collates the reports for CALLBACK, a publicly available monthly newsletter that gives readers both an illuminating and entertaining summary of mistakes, errors, and slip-ups. It also publishes the number of reports received each month. Once again, it is important to stress that no one is named and shamed in the newsletter, nor is it a stage for public self-criticism. The process of reporting errors is referred to by NASA simply as "professional behavior." Any editorial notes are limited to comments on how the content has been compiled – that and no more. There is hardly a better way to facilitate this type of reporting, determine the tone of the reports, and simultaneously ascertain the extent to which CRM is implemented.

Let us take a look at a sample case reported via CHIRP. Because of the need for anonymity, it contains neither names nor the flight number. All we know is that it deals with a large commercial aircraft en route to Tribhuvan International Airport in Kathmandu back in 1988.[203] There were three crew members in the cockpit: the captain, copilot, and flight engineer. The captain was at the controls. The CHIRP documents do not disclose whether it was a passenger or cargo plane.

---

203 CHIRP Feedback (1989), pp. 2–3.

The approach to Tribhuvan is a tricky one. It is impossible to use a conventional instrument landing system with a constant approach angle: the peaks of the Mahabharat Range (up to 9,000 feet high) lie around eight miles south of the landing runway. There is also a mountain crest of nearly 7,000 feet directly in the approach path, and the planes have to fly over it at an altitude of at least 8,200 feet. After that, planes enter a staggered descent to reach the airport, which is itself 4,400 feet above sea level. Five planes smashed into the mountains between 1992 and 2011 alone[204] (Figure 3.13).

The copilot, who submitted the report, remembered the conversation in the cockpit as follows:

Captain: "Descent checklist." The copilot ran through the checklist for the descent together with the captain and flight engineer.

Copilot to Tower: "How's your weather?"

Tower: "Wind zero nine zero degrees, ten knots, visibility five thousand meter, clouds broken at two thousand feet."

Captain: "Okay."

Copilot: "We lost the VOR DME."[205]

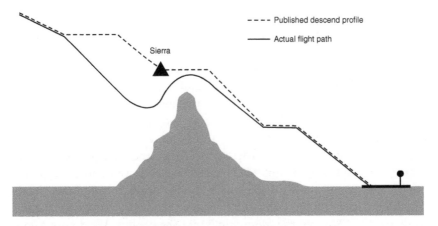

Figure 3.13 **Approach to Tribhuvan over the Mahabharat Range**

---

204 There were two crashes caused by planes colliding with the mountains in 1992 alone. In July, a total of 113 people were killed when an Airbus A310 operated by Thai Airways International crashed. In September of the same year, 167 people lost their lives in a crash involving an Airbus A300B4 operated by Pakistani airline PIA. In both cases, the accidents were caused by navigational errors made by the crew.

205 DME: distance measuring equipment.

Captain: "I am using the Doppler DME."[206]

Copilot: "That's not approved."

Captain: "It's okay. I've been coming to this airport for the last twenty years – since the turbo. Used to come overhead and make a spiral descent."

Copilot: "We are below the MEA [minimum en route altitude]. The mountain ahead is shielding the VOR DME."

Captain: "You worry too much."

Copilot: "Kathmandu Tower, please confirm that your VOR DME is serviceable."

Tower: "Affirmative, you are cleared for approach."

Copilot: "Roger, and for your information we have flags[207] on both VOR and DME."

Tower: "I confirm, both are operative."

Copilot to captain: "You are low, the mountain ahead is shielding the VOR DME signal."

Captain: "No way."

Small breaks in the clouds allowed the copilot to make out a river far below. He was sure this was the Bagmati River that snakes along at the foot of the Mahabharat Range. The fact that the river was in their approach path dispelled any lingering doubts he may have had. He also realized that they would collide with the mountains unless the plane climbed higher. "For God's sake," he said. "Go up to a safe altitude." The captain may have felt some doubts by this point, but he did not react. The copilot initiated the ascent, ordered the flight engineer to apply maximum thrust and said, "Climb to thirteen thousand five hundred, captain. Now." The captain pulled the plane up and it broke through the cloud. The Mahabharat Range was directly below them, they had just managed to fly over it. "You see the mountain," said the copilot. We do not know whether the captain responded to this, but after landing he said "thank you" and "sorry."

---

206 The captain wants to use only the DME rather than the stipulated combined VOR DME signal. Although this gives him information about the distance to the airport, it does not provide details of the course he needs to fly.

207 Red flags on cockpit instruments indicate unreliable information, in this case inadequate signal reception from the VOR and DME that must not be used for navigation.

Reports like this are mainly there to encourage copilots to step in rather than keep silent like Roberti or Gergin on flights ANE 248 and ALW 301, or resign themselves to their captains' decisions like Meurs on KLM flight 4805. Just as illuminating are the reports issued by captains, or captains and copilots together, since by openly describing their errors and mistakes, they turn them from being something unmentionable and embarrassing into something tangible and comprehensible. We will take a look at one such case here, one of those misunderstandings that happens to us all: a situation involving two people, each of whom mistakenly believes that the other agrees with their view on a specific issue. "Shortly after landing, a (twin engine) B777 had to apply heavy braking to avoid overtaking a slower aircraft during taxi to the gate. The Boeing Captain stated to the First Officer that they should 'probably shut one [engine] down' to avoid repeated braking for the slower aircraft. 'I looked over and noticed he [the captain] was guarding the left fuel control and so I shut the right engine down. To our dismay, it soon got very dark and quiet and we realized we had both shut an engine down. The Captain braked the aircraft to a stop. I cranked the APU,[208] remained seated, notified Tower [...] and waited [...] for the resumption of electrical power'."[209]

Generally speaking, it would be interesting to find out how often presumed agreement between two people leads to errors, near misses, and actual accidents. In the crux of these cases, it does not – basically cannot – occur to either party to exchange their views with the other. Another example similar to the one above appeared in the October 2010 edition of CALLBACK. The report was submitted by a captain. "During the cruise phase of our flight we were given a crossing restriction based on the VOR. The instructions were to cross a DME fix[210] south of the VOR at 26,000 feet. The First Officer was flying. When he began the descent, he accidentally depressed the Take Off and Go-Around [TOGA] button which is located in the thrust lever. This resulted in the autopilot being disconnected. The First Officer heard the autopilot disconnect the alarm, but thought that I had disconnected the autopilot. [...] I reached up and silenced the alarm and he interpreted the autopilot disconnect and me silencing the alarm as me having taken over the controls. I never announced that I had taken control of the aircraft. [...] I did not realize he was no longer flying until we reached our assigned altitude and I instructed the First Officer to level

---

208 APU: auxiliary power unit; provides electrical power if the engines are not running.
209 CALLBACK (2008).
210 A navigation point defined in terms of the distance to a radio beacon.

off, and he didn't. I assumed the controls, however we were already through our assigned altitude."[211]

It is a seemingly small episode, yet it illustrates again what we have seen in most of our cases, namely that we better check and cross-check one time too often rather than too rarely – be it in a cockpit or during other forms of cooperation – and that our assumptions very often have a good chance of being wrong.

Another benefit of CRM reporting is that the people who evaluate the reports can look beyond individual errors to identify clusters of problems and address these in subsequent training. We touched on one such phenomenon in the previous example, where captain and copilot lost track of who was controlling the plane and each assumed the other, or the autopilot, was flying the plane. In August 2011, CALLBACK published a special edition on this topic including a variety of personal experiences. Here is another example.[212] "Although I checked and briefed the departure, I failed to notice that the initial heading was incorrect for the current runway's departure (80 degrees difference). [...] Despite briefing until we're blue in the face, somehow a heading that we've both flown a thousand times gets thrown out of our aviation brains due to too much focus on automation. [...] Despite my feeling that something wasn't right, I didn't figure it out until ATC asked us where we were going. This is what drives me nuts about this love affair with automation. No one's flying the plane anymore."

The following example from one of the New York air traffic controllers is particularly nice.[213] "[An aircraft] departing JFK climbing to 11,000 feet [had been] given direct BETTE[214] from Departure Control and handed off to Center. The aircraft checked in climbing to 11,000 feet. I issued [a clearance to] 14,000 feet. The aircraft made approximately a 90-degree turn during climb. When I asked where they were going, the pilot said, oops, they had been cleared direct BETTE, but that they had typed in BETTY which caused the drastic turn off course. Further investigation on the BETTY intersection came up with [coordinates] in the vicinity of Taiwan."

Again, we can identify with what happened here, I assume. How often do we read the first part of a word and automatically complete it in our minds – and end up being wrong – because "BETTY" was actually "BETTE."

211 CALLBACK (2010).
212 CALLBACK (2011).
213 Ibid.
214 Navigation point – identified by five letters according to aviation conventions.

Our next example comes from a captain's perspective. It is typical of the tone of the majority of the reports and illustrates what an excellent device humor can be when describing one's own mistakes.[215] It deals with a night landing where the captain is at the controls and, despite good visibility, mistakenly picks out the wrong landing runway. "So, it's a clear night, calm winds. I've been to [this airport] many times, though never at night. [...] We planned a flaps 40, visual approach to Runway 2L, backed up by the ILS. In retrospect, I should have planned the ILS backed up by the visual, but had I done that, I wouldn't be writing to you. [...] Once I was sure I had the runway, I asked for the visual. We were cleared to descend to 2,000 feet; cleared for the visual approach. [...] I'm already a little low because, like an idiot, I am thinking [the airport] is at sea level, not 400+ feet. So we are out of 2,000 feet, throw the gear down, run the Landing Check, and I am making a slow turn to the runway, or so I thought. Well, I was turning, but looking at Runway 2R. [...] With some help from [the First Officer] I realize I am looking at the wrong runway and make a correction to the left. It wasn't radical as we were still a couple of miles out. It finally dawns on me that I am low so I add power and essentially level off. Now mind you, I've only been flying big jets for about 35 years now and you'd think I'd have realized that if I level off with flaps 40 it will take a lot of power. [...] So to keep it interesting, I add what I think is enough power and of course I'm still concentrating on getting over to the centerline and on the glide slope. [...] Ref[erence] Speed was 132 knots. The First Officer reminded me of that at 130 knots and I pushed the power up a lot as I finally realized that I was essentially going back to the glide slope. [...] I learned this same lesson like 25 years ago, but evidently I left that lesson in some other part of my brain last night. Pass this on: being slow and not on the localizer or glide path at night is very disconcerting. You are completely out of your element and your brain is racing to fix it, but you don't look at the right stuff. An excellent First Officer picks up on that and feeds you the info that you, for whatever reason, aren't getting."

Compared to the cockpit, reports from members of the cabin crew are rare. As a rule, there are around 200–300 ASRS reports from cabin staff per month, whereas there are 10 times as many cockpit reports. However, when CALLBACK focused on "rushing," it featured an ASRS report from a flight attendant who, even with the anonymity clause, showed courage in sharing her error. [216] After all, it could have had fatal consequences. Here, too, the point is that her story should serve as a salutary warning for others: "I felt

---

215 CALLBACK (2012).
216 CALLBACK (2007).

rushed to get the galley in First Class set up. When it was near time for the entry door [to be] closed, I was rushing to put everything away and securing the galley for takeoff. I didn't hear the command to arm the doors for departure. [...] When we reached a certain altitude, the Captain called the purser to notify her that a warning light indicated that door 1R was not closed properly. We looked at the door and then realized that the door was not armed. We tried arming the door, but couldn't. The Captain then descended and proceeded to burn fuel for landing. [...] When we got to a lower altitude, the purser got a call from the Captain saying that the warning light had gone out and we tried arming the door again. This time we were able to do so. We landed ... safely and the mechanics I spoke to determined that the door was not closed properly. They inspected the door, we refueled and left."

To understand the full extent of the "no sanctions" policy applied to error reporting in the aviation industry just imagine the consequences of this flight attendant's error in the context of most companies. Rushing about would be seen as a failure in itself, a sign that she lost her head under pressure. For an employee at the level of a flight attendant, neglecting to secure the door would have meant a black mark on her record for wasting fuel alone. She might have resorted to stating that she definitely closed the door properly if only out of fear and in the hope that the matter would go no further. At the very least, she might have looked for excuses to divert the blame from her: "The captain was in a rush, the purser was hassling me, the first-class passengers were getting impatient," none of them leading to anything anyone could learn something from. What a contrast to the benefits of the calm, intelligent alternative approach, where the problem is identified, the proper measures are introduced to resolve it, and the incident is used to alert others to the risks involved when vital work is forgotten in a hurry. If the door had sprung open, the results would have been far more catastrophic than the mere loss of a load of fuel worth €10,000 – just think back to the case of UAL flight 811. The same applies to pilots who abort and have to repeat an approach because they came in too fast or misjudged a cross wind. They, too, are not required to justify themselves – even if the repeated approach costs a four-figure sum and annoys passengers who are left with little time to make their connecting flights. Any pilot can misjudge an approach. The costs incurred and frustration caused are nothing compared to what would happen if the plane collided with another aircraft or houses near the airport after overshooting the runway.

In addition to the reports submitted to national aviation authorities, the launch of CRM also introduced a debriefing at the end of each flying day, which was to be attended by both cockpit and cabin crew. Depending on the circumstances, it can last anywhere from 5 to 30 minutes. The crew discusses what went well during the flight and what went less well, or even badly. The debriefings are the usual channel for talking through and dealing with any issues. By comparison, the reports and notifications dealt with above are more the exception than the rule.

If a debriefing reveals an irresolvable conflict (i.e., severe negligence or a risky decision that the "guilty party" denies) or the behavior is repeated and becomes a safety hazard, the case can be referred directly to the next level within the relevant airline. In most airlines, that would be the flight safety manager – a role occupied by captains in active service. As a rule, they would first speak with everyone involved, make their judgment, and possibly instruct the relevant parties to undergo additional flight training. As a last resort, they might initiate suspension proceedings. To do that, however, there would have to be a number of similar complaints requiring verification – not only by the safety manager but also by the chief pilot and employee representatives.

Nonetheless, all of these safety measures – and, indeed, CRM error management as a whole – can only work if they are fully accepted, represented, and supported by the individuals involved and their organizations. Otherwise, we end up with cases like Czar 52, which we will look at now.

Czar 52: the limits of CRM

As we have seen, high-level aviation authorities were a driving force in setting up and implementing CRM. The slow-but-sure success of Crew Resource Management validated both them and the people who had worked on the concept. Not only did the number of accidents in the aviation sector fall,[217] but the proportion of pilot errors involving reportable incidents also dropped by nearly 70 percent between 1983 and 2002.[218] In addition, modern behavioral and working models replaced the outmoded hierarchies within the crews. Nonetheless, it is obvious that, for CRM to work properly, the right conditions have to be in place in every organization in which it is used. This also means that it is not just crew members who have to be willing to adhere to CRM. It is also the relevant airlines or, more generally speaking, the organizations behind all the parties involved. The following description of the crash of Czar 52 in 1994 illustrates how vital it is to properly embed CRM in solid organizational structures.

217 Boeing (2006), p. 18.
218 Baker, S.P. et al. (2008), p. 4.

Like the civil aviation sector, the US Air Force had been involved in the development and implementation of the CRM concept from the very start. This is impressive, given that the military has a more complex hierarchy than its civilian counterparts. Its behavioral models, too, were – and still are – considerably stricter than in civil aviation. In such a regimented environment – with its focus on precision and perfection, its distinct sense of rank, and its rules of obedience – it can be quite difficult to talk about errors, either among colleagues or bottom-up.

Even so, the US Air Force was ahead of Britain's Royal Air Force and the German Air Force in getting involved in the development of CRM. Recognizing the validity of its results, it introduced CRM as its behavioral model as early as 1980. Even though rank does not play quite as dominant a role in the Air Force as is the case in the land forces, a superior's orders are clearly not questioned, much less contradicted. This is especially the case in combat situations. However, as in the civil aviation sector, the military realized that following an incorrect order was more serious than raising properly founded concerns. This was based on the same reasoning used in all other organizations that adopted the concept. They wanted to lower the number of errors – no matter if making mistakes fit their image or not – and shape cooperation on board in such a way as to achieve this goal. The smooth cooperation between the team members will have been particularly desirable, as there can be up to six crew members working together in the cockpits of transport planes and bombers.[219] Even fighter jets flown by a single pilot normally fly in pairs or in groups of four or more planes requiring a considerable amount of coordination.

The accident involving Czar 52 occurred 14 years after the introduction of CRM training. Tony Kern has studied this case very closely.[220] My remarks are based on his findings and the official investigation report by the US Air Force.[221]

At 1:58 p.m. on June 24, 1994, a B-52H Stratofortress bomber was taxiing toward takeoff at Fairchild Air Force Base (AFB) near Spokane, Washington. The plane was part of the 325th Bomb Squadron of the 92nd Bomb Wing. Its call sign was "Czar 52." The B-52 is a heavy bomber with eight engines, a wingspan of more than 50 meters, and a maximum takeoff weight of a good 220 tons. According to the US Air Force, a B-52 costs

---

219 The normal crew of a B-52 consists of pilot, copilot, radar navigator, navigator, and electronic warfare officer. An additional crew member may be added depending on the mission.
220 Kern, T. (1995).
221 US Air Force (1995).

$53 million.[222] Even though the B-52 first flew back in 1954, it continues to form the backbone of the United States' bomber fleet (Figure 3.14).

There were four crew members on board Czar 52. The commander was Lt. Col. Arthur "Bud" Holland (46). He had more than 5,000 hours in various B-52 types under his belt. At that time, some of his colleagues regarded him as one of the best B-52 pilots of his generation.[223] He was also a B-52 instructor pilot and head of the 92nd Bomb Wing Standardization and Evaluation Section.[224]

The copilot was Lt. Col. Mark McGeehan (38). With nearly 3,200 flying hours in the B-52, he, too, was an experienced pilot and flight instructor. He was also commander of the 325th Bomb Squadron, part of the 92nd Bomb Wing.

The position of radar navigator/weapon systems officer was filled by Lt. Col. Kenneth Huston (41). Also on board was safety observer Col. Robert Wolff (46). This was to be his farewell flight after an active Air Force career.

Figure 3.14   **A US Air Force B-52H**

---

222 Available at: http://www.acc.af.mil/library/factsheets/factsheet.asp?id=2339 (accessed on August 3, 2012).

223 For example, Col. Michael Ruotsala, former commander of the 92nd Bomb Wing; US Air Force (1995) V-6.4.

224 Lt. Col. Holland was thus responsible for flight instruction, but also for evaluating the flying performance of the squadron's pilots.

Wolff's colleagues, his wife, and two of his sons had already gathered on the airfield below. There was to be a champagne reception in his honor after landing.

The flight was a practice for an airshow scheduled to take place at the AFB on June 26. The maneuvers had been discussed on June 15 in the presence of Col. William Brooks, commander of the 92nd Bomb Wing. The rules stated that the maximum bank of the plane was not to exceed 45°, the maximum angle of attack was to be 25°, and there was to be no formation flying.

However, during the discussion it was also agreed that an exception would be made to allow the flight display to include a "wingover" maneuver. This involves the plane climbing at an angle of around 45° before flying a 180° curve, with a bank of more than 80° permitted for a brief period (Figure 3.15). Although it is possible to execute this spectacular maneuver in the clumsy-looking B-52, a technical order issued by the US Air Force[225] states that it should not be done. The danger of this military aircraft being damaged during the process was seen as too great. There is also a risk that control surfaces could become detached and cause a crash. Even though the wingover clearly breached Air Force guidelines, Col. Brooks still approved the flight plan.

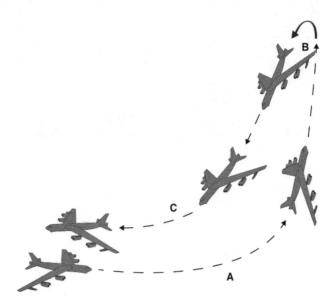

Figure 3.15 **Wingover maneuver**

---

225 USAF Technical Order 1B-52G-1-11.

On June 24, the day of the accident, the crew of Czar 52 was not alone in wanting to practice. The crew of a four-engine KC-135R aerial refueling aircraft was also planning to train for the airshow. Its call sign was "Earl 01." Two weeks earlier, Bud Holland had suggested that the two planes fly in formation during the show. Col. Brooks and the commander of Earl 01 rejected this proposal. Holland defied them both and submitted a flight plan that included a formation flight with Earl 01. For their part, the crew of Earl 01 was planning a solo flight with several touch-and-go maneuvers, where the plane touches down on the runway briefly before immediately taking off again. As planned, Earl 01 initially took off independently of Czar 52.

Czar 52 took off at 1:58 p.m.[226] and Bud Holland executed his maneuvers. However, he did so with much more risk than planned or permitted. At 2:07 p.m., he was hot on the heels of Earl 01. In addition to Czar 52 and Earl 01, there were two other planes in Fairchild Tower's airspace at this point: a C-130 with call sign "Pinto 21" and a UH-1 helicopter with the call sign "Blade 13."

Fairchild Tower [to Czar]: "Five two heavy, traffic is a UH-1, three miles northeast of the field this time, inbound South Training Area. He'll remain well southeast of the runway."

McGeehan: "Five two copies that. And Tower, zero five two heavy, we're left closed, touch and go."

Fairchild Tower: "Czar five two heavy, left closed traffic approved. Company traffic is on five mile dog-leg."

McGeehan: "Zero one got you in sight, we're heading at you. Turning to set them up."

Earl 01: "Okay."

McGeehan: "Turning behind you now."

Fairchild Tower: "Blade one three, sir, if you could transition over to helipad five until these guys have done their demonstration practice."

Holland: "Hey, zero one, we're rolling out on final behind you."

Fairchild Tower: "Earl zero one heavy, will this be your last approach?"

Earl 01: "That's affirmative sir, after this we'll pitch up for the touch and go. Then we're gonna full stop."

---

226 All times are local time (Pacific Daylight Time).

Fairchild Tower: "Roger. Czar five two, will that be the same for you, one more approach, touch and go to a full stop?"

Holland: "That's negative, we'll do a touch and go and stay with you for about another thirty minutes or so."

Fairchild Tower: "Czar five two heavy, roger."

Holland: "Lookin' good here, [Earl] zero one."

Earl 01: "Thanks."

Holland: "I'm off to your five o'clock position. Obviously, I'll just wait 'til you turn in front of me."

Fairchild Tower: "Earl zero one heavy, confirm full stop [landing]."

Earl 01: "Touch and go for Earl zero one."

Fairchild Tower: "Roger."

Holland: "Earl zero one, we'll go by on your left."

Earl 01: "Roger."

Holland: "And Tower, five two, we've got zero one in sight, we'll just adjust."

Fairchild Tower: "Roger."

Holland: "How 'bout just a three sixty [degree] around you right now to get us some spacing?"

Holland had flown over the runway at a height of 250 feet and now started to fly a full circle in preparation for continuing the approach to landing. Unlike in the previous maneuvers, he flew this curve with an initial bank angle of 64°, more than twice as steep as was permitted and far in excess of the 45° Brooks had approved for the airshow. This turning flight should not be confused with the wingover maneuver described previously. In the latter, the plane only maintains the extreme bank angle for a brief moment.

Holland, on the other hand, flew the 360° turn with a consistently high bank angle. Furthermore, he did not maintain the already excessive 64° angle but in fact increased it to 72°. At that moment, the B-52 suffered a partial stall and lost altitude. Holland reduced the bank to 45° and managed to keep control of the plane. However, the distance between the ground and the tips of the wings was now only 100 feet. Despite this, Holland increased the bank again. It was now almost 90°. The plane went into a stall and its nose dipped (Figure 3.16). The air traffic controllers and

Figure 3.16   **Czar 52 immediately before the crash**

other spectators watched in disbelief as the lumbering B-52 fell from the sky with one wing pointing straight up into the air, then hit the ground and exploded.

All four crew members were killed, including Col. Wolff, who died in front of his wife and family. It initially looked like a tragic training accident, but there was more to the story than meets the eye.

Three years before, on May 19, 1991, Lt. Col. Holland had also carried out a demonstration flight in a B-52 at a Fairchild AFB airshow. He completed some spectacular fly-pasts right above the airshow grounds. The flight maneuvers[227] demonstrated the agility of the cumbersome-looking B-52, the skills of Holland, and also impressed the spectators. But already back then, Holland's moves breached the internal flight regulations of the Air Force and the technical limits of the B-52. Worse still, Holland had violated a rule laid down by the US aviation authority[228] stating that flight demonstrations at airshows must never be carried out directly over, or in the direction of, the spectators. This rule was tightened

---

227 A video showing the incidents described and the crash is available at http://www.
   youtube.com/watch?v=YQa4PpIkOZU (accessed on January 28, 2009).
228 FAR Part 91.

in the United States after the Ramstein disaster in 1988, in which several fighter jets from the Italian aerobatic team Frecce Tricolori collided in the air during an airshow performance. One of the planes turned into a fireball and plunged into the crowd. Seventy spectators died that day and around 1,000 were injured. The spectators in Fairchild may not have been aware that Holland had breached regulations. Yet the wing commander Col. Weinman, who was present at the time, definitely would have been. Like all of Holland's other past and future superiors, he refrained from instigating any disciplinary measures.

Barely two months after the airshow in May 1991, there was another demonstration of Holland's "flying skills." It took place on July 12, at an event to mark the departure of Lt. Col. Harper, commander of the 325th Bomb Squadron. In the practice flights and during the official farewell ceremony itself, Holland flew extremely aggressive and daring maneuvers, sometimes at a height of 100 feet. These included the infamous wingover. After this, the Assistant Director of Operations (ADO), Col. Capotosti, decided to take action. Among other things, he was concerned about Holland's influence on younger pilots. Holland was given a verbal warning. No other sanctions were implemented.

At the next annual Fairchild airshow in May 1992, Holland flew maneuvers similar to those of the previous year that once again lay outside the permitted flight profile of the B-52. This also went unchallenged. Naturally, the wingover was part of the program. None of his superiors who were watching the show initiated disciplinary measures. Only Col. Capotosti – who was not present at the time – took action once he heard about the maneuvers. He gave Holland a verbal warning and told him that, if he violated regulations again, he would be banned from flying. In his role as ADO, Col. Capotosti was unable to instigate any further disciplinary measures.

The following year, Holland was tasked with commanding two B-52s as part of a bombing training mission at Farallon de Medinilla Target Range, Guam. Here, too, he drew attention due to his reckless maneuvers. During one exercise, he flew dangerously close to another B-52 while in formation to take photos. In another, he got a crew member to climb into the open bomb bays to film the drops. Both these stunts were in blatant breach of the safety regulations laid down by US Air Combat Command (ACC). After the maneuvers were over, Lt. Col. Bullock, commander of the 325th Bomb Squadron, found out about the photos and videos and attempted to launch formal disciplinary measures against Holland. This never came to fruition. The reasons why remain unclear.

At the Fairchild airshow in August 1993, Holland once again executed maneuvers that flew in the face of the rules. In addition to extremely tight turns and low-level flying, he also performed one maneuver with an excessively steep angle of attack. This meant that the plane ascended practically vertically. Like the wingover, this can lead to material damage, to say nothing of the danger involved.

In March 1994, Holland had the command of a B-52 to demonstrate bomb drops to a camera team from TV station Channel Four at the Yakima training center in Washington State.[229] Like Holland, copilot Capt. Eric Jones was an experienced pilot and B-52 instructor. There were also four other officers on board: Lt. Steve Hollis as a second pilot, Capt. Neil Bannock for electronic countermeasures, and the two radar navigators, Capt. David Laur and Lt. Chris Jorgensson. On this flight, too, Holland performed mostly daring feats. Among other things, he flew so low over the camera team that they had to run for cover. It was only thanks to Laur and Jorgensson that the bomb bays were not open at the time. During the second low-level pass, Holland performed his beloved wingover with a bank that briefly reached 90°. As if that was not enough, he then insisted on flying in formation with smaller, more agile A-10 ground attack bombers. The pilot in charge of this other group agreed. Holland asked how low he planned to fly. "500 feet" was the answer. "500 feet?" queried Holland before replying, "I am going to be down at about 50." During this maneuver, Holland flew toward the camera team so low that Jones grabbed the yoke and pulled the B-52 up while, below, the camera team panicked and dived for cover. The subsequent investigation into Holland's case revealed that the clearance was about three feet. After the event, Jones reported the incident to the squadron operations officer, Maj. Don Thompson, and said he would not fly with Holland again, even if it meant the end of his career as a pilot in the Air Force. Thompson responded by saying that several pilots had already made statements against Holland.

Although Maj. Don Thompson was a friend of Holland's, he reported the violations to commander Lt. Col. Mark McGeehan. Both men came to the conclusion that Holland should be stripped of his flight permit. They therefore turned to the wing flight operations officer, Col. Pellerin. Pellerin spoke to Holland and decided to let him off with another verbal warning. He asked Holland to promise that he would stick to regulations in the future. So Thompson and McGeehan had failed in their efforts to intervene. Nonetheless, McGeehan insisted that pilots from his squadron

---

229 US Air Force (1995) V-28.

were no longer to fly with Holland. Instead, he would take on the role of copilot himself, which also meant that in the months to come, two pilots who definitely resented each other would have to fly together.

At the beginning of June 1994, both pilots were selected for the Air Force's annual airshow. Col. Pellerin selected Bud Holland to showcase the B-52 and flatteringly dubbed him "Mr. Airshow." By now, a large number of the officers in the 325th Bomb Squadron were refusing to fly with Holland. As a result, chief navigator Lt. Col. Huston stepped into the breach for the airshow. The copilot was, of course, Lt. Col. McGeehan. During the very first practice runs, Holland was already flying maneuvers that exceeded the limits of the B-52. No official complaints were made or measures taken.

It was no small wonder, then, that Holland stepped things up a notch at the next practice session on June 24. He would go on to embark on the crazy moves that would lead to the B-52 crash just minutes after takeoff, killing Holland and his fellow officers Robert Wolff, Mark McGeehan, and Kenneth Huston.

The initial report by the US Air Force investigation committee was published on August 10, 1994. The subsequent analysis of the Air Force Accident Investigation Board scrutinized the behavior of Holland and his superiors. It appeared in January 1995 and contained 49 witness interviews. Unsurprisingly, most of the blame for the B-52 crash was put on Holland. To a large extent, that was definitely justified. However, Holland's superiors also bore responsibility for the disaster. They knew what he was like and kept putting up with his antics all the same. For us, this latter aspect is actually the most interesting one. It shows that, even though Holland's behavior was thoroughly overt, he was able to continue flouting the rules of an organization as strict as the US Air Force for at least three years. Virtually nothing was done other than submitting complaints and issuing verbal warnings. This was despite the fact that everyone around Holland probably felt that he was a danger to himself and others – from pilot colleagues to airshow spectators. Was he given free rein to do what he liked again and again simply because his superiors could not be bothered dealing with the situation? Was he shown leniency because his recklessness had something fascinating about it? Or did it also ultimately come down to a failure in leadership within an organization and/or unit? Let us look at the circumstances more closely.

Lt. Col. Holland was undoubtedly an excellent pilot. He was able to push the B-52 to its limits. Such qualities are an advantage in the military, especially in real combat situations, where peacetime regulations are secondary to the objectives of the military mission. This also means that we are dealing with two contradictory rule systems, whose boundaries obviously became blurred for those involved in the case of Czar 52. Most of us will have experienced such situations, albeit in a different context. If I know, for example, that a specific rule – such as maintaining an altitude of 500 feet – is not incontrovertible and will even be overturned in combat situations, I might develop the sense that it is only half a rule, too weak for me to observe it to the letter. Eventually, I might disregard it altogether. Or think of the coexistence of religious and secular doctrines. Here, too, breaches against one set of rules can be canceled out, or at least watered down, by the other; the contrast between strict religious dietary laws and permissive secular-world eating habits is just one of many examples I could cite. The same applies to all other norm systems that exist in parallel, and within which we operate. We encounter these in people's private lives as well, when parents lay down contradictory rules for their children. Without clear orientation, we all – adults and children alike – prefer the system that allows us to do what we want – just as Holland used the existing gray area to do what he wanted. Maybe the people who ought to have intervened felt they could turn a blind eye, without either thinking about the consequences or the double standards that they, too, were applying – both regarding the rules and the people who had to follow them.

However, the worst that one can do when someone constantly breaks the rules is to reward them. In Holland's case, it meant entrusting him with standardization, evaluation, and the flight instruction of pilots in the wing. Together with the freedom he was granted and enjoyed, he must have learned that his behavior was acceptable, at least as long as he did not cause any accidents. There is a small excuse, namely that Holland's development was possible due to permanent changes in leadership on the airbase. This meant the full extent, and the entire sequence, of his violations never really came to light – or only did so after it was too late. Hence, Holland never received retraining, nor did he suffer disciplinary consequences. His suspension from flying, as proposed by Thompson and McGeehan, went unheard. One thing is clear, though: Holland's superiors were aware of CRM. After all, there were already two documented cases in which student pilots taught by Holland had displayed exactly the same disregard for the rules about flying demonstration flights as their teacher. In their cases the respective superiors initiated immediate disciplinary measures and sent them for further training sessions.

What this means is that standards of behavior and leadership principles are of no use to anyone if the organization in question cannot, or will not, ensure they are visibly and tangibly embedded at all hierarchical levels. Successful CRM thus requires a functioning organizational framework in which CRM-compliant behavior is the norm – *accepted and practiced* by everyone. As we have seen from our cases, this process will not – and cannot – happen overnight. In the airline industry, it took 10 years before CRM was designed and established; people had to be convinced of its merits, and the necessary framework in the form of briefings, debriefings, trainings, and information systems had to be installed and used. If this does not happen, a concept such as error management will remain nothing but an idea.[230]

---

230 This is significant given that CRM knowledge and CRM-compliant behavior was checked during the annual evaluations of commercial pilots. However, in contrast to the checks on actual flying skills, pilots cannot be stripped of their flight permits on the basis of an inadequate CRM performance alone. To this extent, an effective CRM concept requires a high degree of intrinsic motivation. Although this is present among the vast majority of flight staff, it still needs to be implemented more consistently.

Part IV Error Management

At the end of this book, we are left with the question of how CRM, which has proven to be so successful in aviation, can be transferred to other sectors and implemented in everyday business life. Or, to go back a step: to whom does this form of error management apply aside from those in the aviation industry? After all, unlike most other industries, aviation is a risk industry and although every industry or company has its own particular risk areas, managers do not arrive at work knowing that they are responsible for the safe transport of hundreds of people each day. They are, however, in charge of business processes, the success of their particular division, and for keeping their work force employed. So the number of errors they make should be limited as well. From this perspective, the answer to the question above is simple: error management is relevant to every organization that wishes to reduce error volumes, whether or not it is in a risk industry. In fact, most organizations will already have taken steps in this direction by trying to eliminate potential error sources and attempting to analyze and resolve errors that have occurred.

But why then set up a new error management system that is based on the structure of a different industry? If you have read parts I to III, you will know the answer and realize the fundamental difference between the traditional approaches to preventing errors and the error management used in CRM. To sum it up: in conventional approaches, errors are stigmatized as individual weaknesses, whereas modern error management accepts them as an unavoidable aspect of human behavior. While both strategies seek to avoid errors, the former puts them in a negative light and associates them with embarrassment, shame, fear, and punishment. In the latter, those who have made the errors might become annoyed at themselves, but they need not fear any sanctions. Instead, they – possibly together with others – will analyze what led to the mistake and eliminate this so that it does not cause any future problems. This also means that the neurotic mindset generated by the conventional fear of mistakes is replaced by factual investigations. Van Dyck et al. explain it as follows: "One way to contain the negative and to promote the positive consequences of errors is to use error management. [...] the error management approach distinguishes between errors and their consequences. Whereas error prevention aims at avoiding negative error consequences by avoiding the error altogether, error management focuses on reducing negative error consequences and on increasing potentially positive consequences. [...] In addition, error management ensures that errors are quickly detected and reported, that negative error consequences are [...] minimized, and that learning occurs."[231]

So far, so good. Let us turn to another question, one that is much more difficult to answer, namely: how do we actually implement error management? Thankfully, the aviation industry has led the way here and tested and established the mechanisms we will examine below. The only condition that cannot easily be replicated is the internal mindset needed for this endeavor. Among other things, it requires us to give up our normal ways of thinking and behaving, which is easier said than done, as it means replacing old habits with new ones and varying or revising familiar thought patterns. Although we make this type of mental shift in other situations all the time, we normally do it unconsciously and fluidly. Still, it rarely happens overnight. Just think of our ideologies, values, convictions, opinions, wishes, preferences, dislikes, and the feelings associated with them. We change these based on trends, political situations, new findings, advertising, and opinions expressed in the media. Sometimes we even do it consciously and give up destructive behaviors, either through our own willpower or with professional help.

So, if it is possible to change ingrained habits, we should also be able to alter our attitude toward errors, right? The difficulty is that we have internalized this attitude since childhood and it has led to "keystone habits," as Charles Duhigg[232] terms them in *The Power of Habit*, meaning the habits that are the hardest to change. In his book he shows that it is possible to alter them if we understand how they *work*.[233] I would like to add here that we first have to understand where they *come from*. If I know that I acquired a particular habit in my younger years and repeated it until it became part of my value system and my view of life, then I also know that I will need time to unlearn it and ultimately free myself from it. In this respect, understanding how my habit works is not of much use to me. To take an example: if I, as a child and member of group B, am taught to despise the members of group A primarily because they belong to group A, it may be that as an

---

231 Dyck, C. v. et al. (2005), pp. 1228–1229.
232 Duhigg, C. (2012), p. xvi.
233 Ibid.

adult I will grasp the potentially dangerous and destructive nature of this notion. I understand how my habit *works*. I understand that it harms other people and perhaps even me. I take steps to change it because I realize that it is irrational and detrimental. Then, finally, I succeed in overcoming it. Mentally, that is. Then, after one single controversial encounter with a member of group A, I find myself once again consumed by contempt, as I have failed to get to the roots of my emotions and weed them out. Depending on my perseverance, I either renew my efforts to work on the surface behavioral controls, or revisit my deep-seated psychological patterns and start the long process of restructuring them.

Our attitude to errors is not quite so charged and we might not need a psychological analysis to alter it. To describe the required change process in a simplified form, I refer to Duhigg, who compares it to a loop made up of three stages. "First, there is a *cue*, a trigger that tells your brain to go into automatic mode and which habit to use. Then there is the *routine*, which can be physical or mental or emotional. Finally, there is a *reward*, which helps your brain figure out if this particular loop is worth remembering for the future."[234]

Here, too, I would like to add that the third element can equally be *punishment*, as our brains register punishment as well as reward. Traditionally, the sequence for the one who made the mistake would therefore be:

Cue = error → *routine* = *dismay/embarrassment/shame/fear* → *reaction* = *denial or confession.*

For the person noting someone else's mistake, it would be:

Cue = error → *routine* = *anger* → *punishment* = *sanction*; or: cue = error → *routine* = *silence* → *punishment* = *silent resentment.*

Actually, Duhigg shows the situation after the successful introduction of CRM, namely:

Cue = error → *routine* = *analysis/processing* → *reward* = *new knowledge and confidence that the error will be avoided in the future* (Figure 4.1).

Error management thus begins with a new mindset, one that knows errors are normal and have to be accepted. As far as organizational error management is concerned, it should be clear that even though the implementation has to be

---

234 Ibid., p. 19.

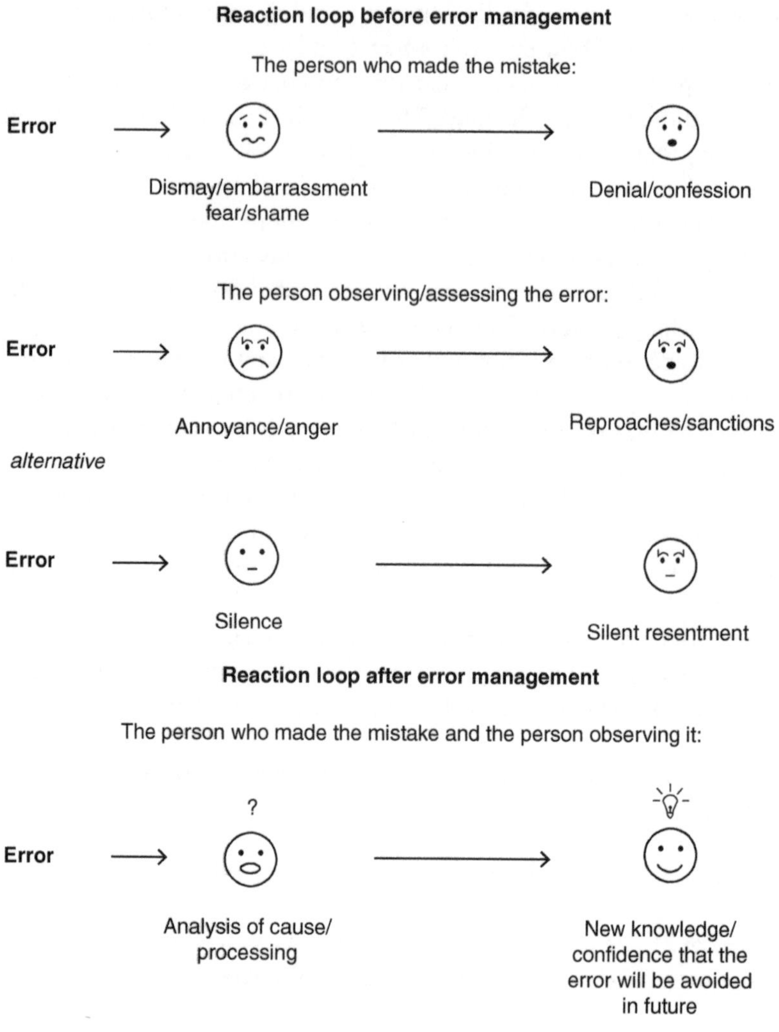

Reaction loop before error management

The person who made the mistake:

Error ⟶ 😟   ⟶ 😮

Dismay/embarrassment   Denial/confession
fear/shame

The person observing/assessing the error:

Error ⟶ ☹️   ⟶ 😮

Annoyance/anger   Reproaches/sanctions

alternative

Error ⟶ 😐   ⟶ ☹️

Silence   Silent resentment

Reaction loop after error management

The person who made the mistake and the person observing it:

?   💡

Error ⟶ 🙂   ⟶ 😊

Analysis of cause/   New knowledge/
processing   confidence that the
                                        error will be avoided
                                        in future

Figure 4.1   **Reaction loop before and after error management**

a management decision, the overall success will depend on individuals. Role
models have to be those people with administrative authority, since they are
the ones who can guarantee freedom from sanctions if errors occur.

Dealing with errors as a team/group means openly and calmly addressing and discussing errors. It is another routine that will need time, but above
all, it requires that the teams/groups stay with it even when it becomes clear
that changing their behavior requires patience. Remember that it was 10 years

before CRM became a fixture in the leading airlines. After another 10 years it had become a global standard – which gives you an idea of the efforts involved in the process we are talking about. For this reason alone, it would be advisable to obtain support, both on the individual and group levels, from external coaches who can observe and advise during the initial implementation phase.

Prior to this, the leadership style and/or team behavior of all involved needs to be evaluated with regard to the requirements of error management, and the communication within the company has to be analyzed to see how open or closed it is so that everybody can work from there. It is certainly not necessary to initiate a huge turnaround process and plan major roll-outs. At the beginning, it is enough to concentrate on central cooperation interfaces and start applying error management at these points, either in departments, project groups, or a structure across the company. Here, too, employees should discuss the results of the first evaluations with coaches and see how the latter can assist with behavioral modifications.

Operational weaknesses and the influence of external factors on processes may also be identified in this way, provided that the individual processes are documented as they are in the aviation sector. "In addition to making changes prompted by crashes, NTSB recommendations, and FAA regulations, airlines have created databases that analyze flights, identifying threats to safety such as adverse weather (the most common threat), radio congestion, and operational pressures, and indicating whether the flight crews identified and managed the threats. The results help airlines to identify and address problems within their operations."[235]

As was the case with captains, the greatest resistance to introspection will probably come from those in senior management positions. According to a study by Guillén et al. (2012), this attitude has emerged during general evaluations of leadership styles. "People who are highly motivated to lead because leadership is a part of their self-concept may be less susceptible to assess their own skills when evaluating their probability to engage in managerial duties. As our results suggest, people who value the leadership role may be motivated to lead even with low levels of competence. Therefore, highlighting the importance for people to understand the appropriate skills and knowledge for adopting new leadership responsibilities, as well as having the chance to calibrate their own skills and knowledge against them, can be very valuable."[236]

Guillèn et al. see similar problems with managers' abilities to be team players. "Although the value of being a good team player and an affiliative

---

235 Baker, S.P. et al. (2008), pp. 4–5.
236 Guillén, L. et al. (2012), p. 30.

leader is often highlighted in leadership theory and leadership development practice, the managers that are reporting higher levels of affective
motivation to lead are those who do not score high on 'affiliation.' This
may indicate that even if leadership theory has evolved toward a more
communal understanding of the leadership role [...], managers motivated to
lead do not necessarily score high in collaborative values. This mismatch indi-
cates that the development of relationships and collaborative skills is crucial
in leadership development efforts of organizations."[237] In other words, managers may need to make as much effort adjusting to the openness required in
error management as did the flight captains at the end of the last century. As
I mentioned in part I, we are dealing with the modifications of a traditional
self-image – moreover an image exaggerated to mythical dimensions – which
means that parting from it will necessarily be painful. However, since managers do make mistakes, they will have to practice modern methods of error
management to deal with them. As captain Al Haynes said in retrospect:
"Up until 1980, we kind of worked on the concept that the captain was
THE authority on the aircraft. What he says, goes. And we lost a few airplanes because of that. And we would listen to him and do what he said and
wouldn't know what he is talking about. [...] The days of the captain being
the ultimate authority are gone. He may be the authority on the airplane,
he may sign for the papers and all this, but you don't work that way."[238] Let
me stress again that error management does not countermand the authority
of the captain or manager. It simply levels the hierarchy to the extent that
people can communicate with their superiors openly and without fear.

Seminars that could accompany this path toward error management
should be freshly designed with the emphasis on practical exercises. Instead
of offering theory and practice in a 50:50 ratio, real-life error scenarios have
to be reenacted and realistic error situations simulated, complete with comprehensive debriefings. One more thing: in the aviation sector, simulator
exercises are carried out every six months, which should serve as a guideline
regarding the frequency and regularity of refresher courses so that specific
reactions and behaviors become routine. In our context, a one-off, one-
week seminar is not enough.

With respect to the team members, the same approach applies
as to managers. In the case of the former, it is not the hierarchical
hurdle that has to be overcome, but rather the identities and roles each
individual has adopted and will not easily give up. To illustrate this, let

---

237 Ibid.
238 Haynes, A. (1991).

me cite the roles defined by Meredith Belbin. According to him, there are three groups, namely: cerebral roles, people-oriented roles, and action-oriented roles. These are divided into the subgroups of plant, monitor/evaluator, specialist (cerebral); coordinator, team worker, resource investigator (people-oriented); shaper, implementer, completer/finisher (action-oriented).[239] At first glance, we may guess that "specialists" find it particularly hard to own up to errors and that non-specialists find it difficult confronting them with these errors; after all, the role of the specialist is based on the assumption of him or her having privileged knowledge. But specialists can get it wrong, too, just like all the others in Belbin's typology, as each role has its weaknesses that make its owner susceptible to errors.

The thing all team members have to learn – as we saw with our copilots and flight engineers – is to be frank and to alert their superiors to errors, ask if one of the top-down decisions is unclear, trust their own judgment, and express their opinions; remember the case of the copilot and captain approaching Tribhuvan International Airport in Kathmandu. Over time and with continuous exercising, it becomes easier for everyone to ask questions, raise objections, and express contrary opinions. This is especially true once the criteria of error management have been organizationally internalized so that everyone understands that it is not about pointing the finger and showing who is wrong and who is right. The only thing that matters is preventing possible mistakes. Just as in the cockpit, all we need is to calmly and objectively alert the person in danger of erring, which ideally will lead to an equally calm "thank you" in response.

Introducing modern error management therefore means entering new territory and determining the appropriate way for your own company, department, or team when moving toward that goal. But basically, all error management processes can initially be modeled on the structure used in the aviation industry.

The first phase may already be complete for companies active in risk sectors. Usually, the task here is to create a safety situation for all involved, and to specify procedural standards along with relevant checklists and related training. In most industries, identifying error-prone interfaces and analyzing communication models is an adequate first step.

The second phase requires listing potential sources of error and supplementing this list continuously. It also means establishing a safeguard system by which the actions of each individual are counterchecked by a second person.

---

239 Belbin, R.M. (2006), p. 22.

In the third phase – the one which interests us most here – errors are discussed in the respective team, without taking account of each individual's position in the hierarchy or their role in the team.

Whether a company decides to adopt an additional, anonymous reporting system similar to CHIRP, ASAP, and ASRS, is up to each one of them. Just keep in mind that CHIRP, ASAP, and ASRS are overarching initiatives supported by either a state or public body to ensure the anonymity of the reports and protect the system from being abused. I think it is doubtful that the same effect can be guaranteed within a company. On the other hand, a system like those mentioned above may not even be feasible, since very few companies have processes that are as standardized and regulated as those in the aviation industry.

Starting an in-house error management newsletter like CALLBACK seems to be a better way to collate error reports, cases, and topics for special editions. At the beginning, this type of reporting will also be difficult to implement. That will only change once it is possible for people to identify with the stories, or when these are controversial. The main thing is that cases, reports, and examples do not serve to laud the author/s in a way no intelligent reader will believe. Contributions from senior management could help to break the ice, particularly if they are written in such a light and humorous tone as the CALLBACK reports we read. That too will take a while.

Ultimately, the figures will speak for the success of error management. To that end, let us look once more to the aviation sector and the results of the already mentioned longitudinal study by Baker et al. (2008): "The NTSB reported 604 accidents involving domestic air carriers in the United States during 1983–2002. We included 558 (92%) in our analysis. Of the remaining 46 mishaps, 39 were excluded due to missing information and 7 for miscellaneous reasons. The rate of mishaps over the 20-yr period averaged 33 per 10 million flights, with no consistent longitudinal trend. [...] Pilot error was noted in 180 mishaps (32%). The percentage of mishaps that involved pilot error declined from 42% in 1983–87 to 36% in 1988–92, 31% in 1993–97, and 25% in 1998–2002 [...], *an overall decrease of 40%* [italics added]. [...] The most common types of pilot error were carelessness (26% of errors), poor decisions (23%), mishandling aircraft kinetics (21%), and poor crew interaction (11%). [...] During the period studied, the rate of mishaps related to pilot error declined by 40%, from 14.2 to 8.5 per 10 million flights [...]. Pilot error categorized as flawed decision declined from 6.2 to 1.8 per 10 million flights, a 71% reduction [...]. Poor crew interaction declined from 2.8 to 0.9 per

10 million flights, a 69% reduction"[240] (Figure 4.2). The percentages that interest us most are the 40 percent drop in pilot errors and the 69 percent drop for the crew.

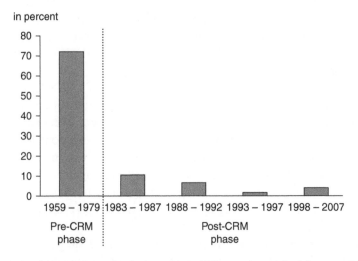

Figure 4.2 **Percentage of pilot errors in US commercial airline accidents before and after the introduction of CRM in 1980**

The breakdown for the pilot errors is as follows (Figure 4.3):

	Total errors *	
Category of pilot error	N	Percent
Careless	62	26.3
Flawed decisions	55	23.3
Mishandled aircraft kinetics	50	21.2
Poor crew interaction	27	11.4
Mishandled wind or runway conditions	17	7.2
Other	25	10.6
Total	236	100.0

* Includes up to two pilot errors per mishap, if in different categories.

Figure 4.3 **Number and percentage of pilot errors, by the type of error, US air carrier, 1983–2002**

---

240 Baker, S.P. et al. (2008), p. 3.

The drops in Figure 4.2 are, of course, not down to CRM alone. The hypothesis proposed by Baker et al. specifies: "The decreasing trend in pilot-error mishaps probably reflects, in part, advances in technology such as cockpit displays that permit pilots to avoid threatening weather and to determine their exact location in relation to airports, runways, etc. Emphasis in recent years on crew resource management has likely been another important factor in the 68% reduction in mishaps involving poor crew coordination."[241]

Likewise, neither CRM, nor error management will be able to make a company more innovative and experimental, contrary to what van Dyck et al. suggest. "In the long run, organizations that have an effective approach to errors may be more profitable because these organizations learn from errors, are more apt to experiment, and are more likely to innovate."[242] In fact, neither CRM nor any other kind of error management needs these justifications and assurances. Setting it up properly is an achievement in itself, as is the personal effort and exertion that are part and parcel of its successful implementation for all involved.

To sum it up in the words of Kathryn Schulz, I suggest that we all remember: "We err because we believe, above all, in ourselves: no matter how often we have gotten things wrong in the past, we evince an abiding and touching faith in our own stories and theories. Traditionally, we are anxious to deny that those stories and theories are stories and theories – that we must rely on our own imperfect representations to make sense of the world, and are therefore destined to err."[243]

Getting it right the next time – or doing it in a better way – is what error management promises us in the event that something goes wrong again this time. That is what I wish for us all, both in our private and in our business lives.

---

241 Ibid., p. 4.
242 Dyck, C.v. et al. (2005), p. 1228.
243 Schulz, K. (2010), pp. 338–339.

Sources for Figures

Part I Pre Crew Resource Management

1.1 Waser, Ch. (2009). De Havilland Canada DHC-6-310 Twin Otter aircraft. http://www.airliners.net/photo/Isles-of-Scilly/ De-Havilland-Canada/1853023/M/. 04.06.2012.

1.2 Floßmann, S. (2012). Berlin, Germany: Artegraph.

1.3 National Transportation Safety Board (1978). Aircraft Accident Report. Japan Air Lines Co., McDonnell-Douglas DC-8-62F, JA 8054, Anchorage, Alaska, January 13, 1977. NTSB-AAR-78-7. Washington, DC: NTSB.

1.4 Plomitzer, G. (1995). Boeing 757–225 aircraft. http://www.airliners. net/photo/Birgenair/Boeing-757-225/1407352/M/. 05.06.2012.

1.5 Floßmann, S. (2012). Berlin, Germany: Artegraph.

1.6 Mahr, L. (2012). Berlin, Germany: Lena Mahr.

1.7 Wright, A. (2006). Boeing 757–251 aircraft. http://www.airliners.net/photo/Northwest-Airlines/Boeing-757-251/1066120/L. 18.06.2012.

1.8 Search, J. (2004). Boeing 757–222 aircraft. http://www.airliners.net/photo/United-Airlines/Boeing-757-222/0568430/L. 08.06.2012.

1.9 Wien, K. (2008). Airspeed indicator in the Boeing 757. http:// www.flickr.com/photos/flyforfun/2596073793/sizes/o/in/photostream/. 04.06. 2012.

1.10 Floßmann, S. (2012). Berlin, Germany: Artegraph.

1.11 Floßmann, S. (2012). Berlin, Germany: Artegraph.

1.12 Mahr, L. (2012). Berlin, Germany: Lena Mahr.

1.13 Central Press/Hulton Archive (1977). Air Disaster. gettyimages.

1.14 Floßmann, S. (2012). Berlin, Germany: Artegraph.

1.15 Chaloner, H. (1978). McDonnell Douglas DC-8-61. http://www. airliners.net/photo/United-Airlines/McDonnell-Douglas-DC-8-61/0554689/L/. 03.04.2012.

1.16 de Groot, P. (1974). Douglas DC-8-33. http://www.airliners. net/photo/Martinair-Holland/Douglas-DC-8-33/0695173/L/. 21.05.2012.

1.17 National Transportation Safety Board (1979). Aircraft Accident Report. United Airlines, Inc., McDonnell-Douglas, DC-8-61, N8082U. Portland, Oregon, December 28, 1978. NTSB-AAR-79-7. Washington, DC: NTSB.

1.18 Karm, B. (1978). UAL 173 after the emergency landing. http:// pdxretro.com/2010/12/united-flight-173-crashed-on-this-date-in-1978/. 06.08.2012.

1.19 Luke, J./Photolibrary. Air traffic controller. gettyimages.

1.20 Garrard, B. (1982). Boeing 707–321. http://www.airliners.net/ photo/Avianca/Boeing-707–321B/1109024/M/. 22.05.2012.

1.21 Floßmann, S. (2012). Berlin, Germany: Artegraph.

1.22 National Transportation Safety Board (1991). Aircraft Accident Report. Avianca, The Airline of Columbia, Boeing 707-321B, HK 2016, Fuel Exhaustion, Cove Neck, New York, January 25, 1990. NTSB/AAR-91/04. Washington, DC: NTSB.

1.23 Hass, P. (1990). AVA 052. http://www.airdisaster.com/photos/avianca52/photo.shtml. 29.09.2012.

Part II   Crew Resource Management

2.1 Robinson, D. (update 2012). Handley Page Type W8 1920. http:// www.aviationancestry.com/Aircraft/HandleyPage/HandleyPage-TypeW8–1920-1.html. 04.06.2012.

2.2 Gantz, C. (update 2010). DC-3. http://www.industrialdesignhistory.com/node/157. 18.07.2012.

2.3 Boeing Commercial Airplanes (1990). Statistical Summary of Commercial Jet Airplane Accidents – Worldwide Operations 1959–1989. Seattle, OR.

2.4 Hawkins, F.H., H.W. Orlady (1992). Human Factors in Flight. Aldershot, Great Britain: Ashgate Publishing.

2.5 Edwards, E. (1975). Stress and the Airline Pilot. Paper presented to British Airline Pilots Association Technical Symposium: Aviation Medicine and the Airline Pilot. London, Great Britain.

2.6 Floßmann, S. (2012). Berlin, Germany: Artegraph.

2.7 Airdisaster.com (1983). Air Canada 797. http://www.airdisaster. com/photos/ac797/2.shtml. 06.08.2012.

2.8 Reason, J. (1997). Managing the Risks of Organizational Accidents. Aldershot, England: Ashgate Publishing.

Part III Post Crew Resource Management

3.1 Chernoff, E.M. (1987). Boeing 747–122. http://www.airliners.net/photo/United-Airlines/Boeing-747-122/0881342/L/. 22.05.2012.

3.2 Floßmann, S. (2012). Berlin, Germany: Artegraph.

3.3 Rutherford, M. (update 2012). United Flight 811. http://www.airdisaster.com/eyewitness/ua811.shtml. 06.08.2012.

3.4 Floßmann, S. (2012). Berlin, Germany: Artegraph.

3.5 National Transportation Safety Board (1990). Aircraft Accident Report. United Airlines Flight 232, McDonnell-Douglas DC-10–10, Sioux City, Iowa, July 19, 1989. NTSB/AAR-90/06. Washington, DC: NTSB.

3.6 Aviation News (update 2012). Pilot Who Helped Fly Crippled Jet From Denver Dies. http://aviationnewsdaily.com/2012/05/09/pilot-who-helped-fly-crippled-jet-from-denver-dies/. 01.11.2012.

3.7 National Transportation Safety Board (1990). Aircraft Accident Report. United Airlines Flight 232, McDonnell-Douglas DC-10–10, Sioux City, Iowa, July 19, 1989. NTSB/AAR-90/06. Washington, DC: NTSB.

3.8 Taro Yamasaki/TIME & LIFE Images (1989). gettyimages.

3.9 Aviation Safety Reporting System – ASRS. http://asrs.arc.nasa.gov/index.html. 01.11.2012.

3.10 Aviation Safety Reporting System – ASRS. http://asrs.arc.nasa.gov/index.html. 01.11.2012.

3.11 Confidential Human Factors Incident Reporting Programme – CHIRP. http://www.chirp.co.uk/reporting-guidelines.asp. 01.11.2012.

3.12 Aviation Safety Reporting System – ASRS (2012). http://asrs.arc.nasa.gov/publications/callback/cb_391.html. 01.11.2012.

3.13 Mahr, L. (2012). Berlin, Germany: Lena Mahr.

3.14 US Air Force. http://www.af.mil/shared/media/photodb/photos/021105-O-9999G-011.jpg. 01.11.2012.

3.15 Mahr, L. (2012). Berlin, Germany: Lena Mahr.

3.16 US Air Force (1995). AFR 110-14 USAF Accident Investigation Board. 24 June 1994, Fairchild AFB, WA, B-52H Aircraft S/N

61–0026 92 BW 325 BS. HQ 12th Air Force. Obtained under the Freedom of Information Act on April 30, 2009 by the Department of the Air Force, Headquarters Air Combat Command. Langley Air Force Base, Virginia.

Part IV Error Management

4.1   Hagen, J. (2012). Berlin, Germany: ESMT CS GmbH.
4.2   Baker, S.P.,Q. Yandong, G.W. Rebok and G. Li (2008). Pilot Error in Air Carrier Mishaps: Longitudinal Trends Among 558 Reports, 1983–2002. Aviation, Space, and Environmental Medicine, 79 (1): 2–6.
4.3   Baker, S.P.,Q. Yandong, G.W. Rebok and G. Li (2008). Pilot Error in Air Carrier Mishaps: Longitudinal Trends Among 558 Reports, 1983–2002. Aviation, Space, and Environmental Medicine, 79 (1): 2–6.

Glossary and Abbreviations

ADI	attitude director indicator – artificial horizon
AFB	air force base
Aileron	control surfaces on the wings to control roll movements – used in combination with the rudder to avoid yawing; ailerons always operate in opposite direction of each other
Angle of attack	angle between the airfoil chord line and the relative wind – critical for determining a stall condition
APU	auxiliary power unit; provides electrical power if the engines of an aircraft are not running
ARTCC	Air Route Traffic Control Center
ASAP	Aviation Safety Action Program
ASRS	Aviation Safety Reporting System
ATC	air traffic control
Captain	pilot licensed to fly a multi-crew aircraft as pilot in command
CEO	chief executive officer
CHIRP	Confidential Human Factors Incident Reporting Programme
Col.	Colonel
Cpt.	Captain
CRM	Crew Resource Management
CVR	cockpit voice recorder
DME	distance measuring equipment – usually in combination with a VOR, provides distance information
EGT	exhaust gas temperature of a jet engine (parameter for managing the power of a jet engine)
EICAS	Engine Indication Crew Alerting System – warning and alert messages display in the cockpit
Elevator	control surface at the tail of an aircraft to control the pitch (up and down movements)

FAA	Federal Aviation Administration
FDR	flight data recorder
First officer	pilot licensed to fly a multi-crew aircraft, but not qualified as pilot in command – also referred as co-pilot
FMC	flight management computer – contains all relevant flight data; in flight all navigational inputs for the autopilot are done via the FMC
Foot	unit for measuring altitude in aviation (equals 0.3048 meters)
Glide slope	usually part of an ILS – provides vertical information of an aircraft in relation to the runway
GPWS	Ground Proximity Warning System
ICASS Group	International Confidential Aviation Safety Systems Group
ILS	Instrument Landing System – a navigation system for landing. It consists of a localizer and glide slope
Knot	unit for measuring speed in aviation (1 nautical mile per hour or 1.852 km per hour)
LNAV	lateral navigation mode of the autopilot
Localizer	usually part of an ILS – provides lateral information of an aircraft in relation to the runway
LOFT	Line Oriented Flight Training
Lt.	Lieutenant
Lt.Col.	Lieutenant Colonel
Maj.	Major
MAYDAY	call to report an emergency via radio
Mile	nautical mile (equals 1.852 km) – unit for measuring distance in aviation
N1	speed of the low pressure rotor of a turbo-fan engine (in percent of the maximum allowable speed; parameter for managing the power of a jet engine)
NASA	National Aeronautics and Space Administration
NTSB	National Transportation Safety Board – US agency responsible for analyzing aviation accidents and compiling accident reports
PA	public address (cabin announcement in an aircraft)
Pilot flying	pilot flying the aircraft – may be the captain or the first officer

Pilot in Command (PIC)	pilot who is legally responsible for the flight – must be a captain on a multi-crew aircraft
Pilot monitoring	non-flying pilot supporting and monitoring the pilot flying – may be the captain or the first officer
Pitch angle	angle between the longitudinal axis of the aircraft and the horizon – displayed on the ADI
Pound	unit for measuring weight in the United States (equals 0.45 kg)
QRH	Quick Reference Handbook – contains all checklists and procedures for normal and non-normal operations of an aircraft (used whenever a problem arises)
Rudder	control surface at the vertical tail of an aircraft to control the direction (left and right movements) – used in combination with ailerons to avoid yawing
RVR	Runway Visual Range – runway visibility measures provided during low visibility conditions
SAM	San Francisco Aero Maintenance (technical support center of United Airlines)
SHEL	software, hardware, environment, liveware
Stall	loss of lift after the angle of attack increases beyond a critical value; if not recovered quickly by reducing the angle of attack – that is, lowering the nose of the aircraft and or increasing power – it results in rapid loss of altitude and may lead to loss of control of the aircraft
TAG	Trans-Cockpit Authority Gradient
UTC	Universal Time Coordinated (former GMT)
V1	maximum speed at which a rejected takeoff can be aborted, in the event of an emergency
V2	minimum speed that needs to be maintained after takeoff up to acceleration altitude, in the event of an engine failure after V1
VNAV	vertical navigation mode of the autopilot
VOR	Very High Frequency Omnidirectional Radio Range – a radio range used for navigation
Vr	rotation speed that ensures lift-off is possible at takeoff

Bibliography

Aviation Safety Reporting System – ASRS (2011). Report ACN 981661, November 2011.

Baker, S.P., Q. Yandong, G.W. Rebok and G. Li (2008). "Pilot Error in Air Carrier Mishaps: Longitudinal Trends Among 558 Reports, 1983–2002", *Aviation, Space, and Environmental Medicine*, 79 (1): 2–6.

Bartelski, J. (2001). *Disasters in the Air*. Shrewsbury, England: Airlife Publishing.

Belbin, M. (2006). *Team Roles at Work*. Oxford, England: Buttlerworth-Heinemann.

Bernard, A., S. Hamilton and D.A. Marchand (2002). "The Barings Collapse", IMD Case Study 1–0155 und 1–0156.

Blake, R.R. and J.S. Mouton (1964). *The Managerial Grid*. Houston, TX: Gulf Publishing Company.

Boeing Commercial Airplanes (1990). *Statistical Summary of Commercial Jet Airplane Accidents – Worldwide Operations 1959–1989*. Seattle, OR.

Boeing Commercial Airplanes (2006). *Statistical Summary of Commercial Jet Airplane Accidents – Worldwide Operations 1959–2006*. Seattle, OR.

Braunburg, R. (1978). *Kranich in der Sonne – Die Geschichte der Lufthansa*. München, Germany: Kindler Verlag.

Burns, T. and G.M. Stalker (1961). *The Management of Innovation*. London, England: Tavistock Publications.

Byrnes, R.E. and R. Black (1993). "Developing and Implementing CRM Programs", in Wiener, E.L., B.G. Kanki and R.L. Helmreich (eds) *Cockpit Resource Management* (421–443). San Diego, CA: Academic Press.

CALLBACK from NASA's Aviation Safety Reporting System (2007). Issue 331, July 2007.

CALLBACK from NASA's Aviation Safety Reporting System (2008). Issue 346, October 2008.

CALLBACK from NASA's Aviation Safety Reporting System (2010). Issue 370, October 2010.

CALLBACK from NASA's Aviation Safety Reporting System (2011). Issue 379, August 2011.

CALLBACK from NASA's Aviation Safety Reporting System (2012). Issue 391, August 2012.

Civil Aviation Authority – Safety Regulation Group (2006). CAP 737, Crew Resource Management (CRM) Training. Issue 2, November 29, 2006.

Comisión de Investigación de Accidentes e Incidentes de Aviación Civil (1979). A-102/1977 y A-103/1977 Accidente Ocurrido el 27 de Marzo de 1977 a las Aeronaves Boeing 747, Matrícula PH-BUF de K.L.M. y Aeronave Boeing 747, matrícula N736PA de PANAM en el Aeropuerto de los Rodeos, Tenerife (Islas Canarias). http://www.fomento.gob.es/MFOM/LANG_CASTELLANO/ORGANOS_COLEGIADOS/CIAIAC/PUBLICACIONES/HISTORICOS/A-102–103–1977/. 26.10.2012.

Confidential Human Factors Incident Reporting Programme – CHIRP (1989). Feedback. Issue No. 20, December 1989.

Conlee, M.C. and A. Tesser (1973). "The Effects of Recipient Desire to Hear on News Transmissions", *Sociometry*, 4: 588–599.

Cooper, G.E., M.D. White and J.K. Lauber (1980). "Resource Management on the Flightdeck: Proceedings of a NASA/Industry Workshop", NASA CP 2120. Moffett Field, CA: NASA-Ames Research Center.

Detert, J.E and A.C. Edmondson (2011). "Implicit Voice Theories: Taken-for-Granted Rules of Self-Censorship at Work", *Academy of Management Journal*, 54 (3): 461–488.

DGAC Press Release (1996). Birgenair B757 Accident, March 18, 1996. (Translation, released by US NTSB). http://www.rvs.uni-bielefeld.de/publications/Incidents/DOCS/ComAndRep/PuertoPlata/18–3-birgenair.html. December 15, 2008.

Duhigg, C. (2012). *The Power of Habit*. London, England: William Heinemann.

Dyck, C. v an, M. Frese, M. Baer and S. Sonnentag (2005). "Organizational Error Management Culture and Its Impact on Performance: A Two-Study Replication", *Journal of Applied Psychology*, 90 (6): 1228–1240.

Edmondson, A.C. (1996). "Learning from Mistakes Is Easier Said Than Done: Group and Organizational Influences on the Detection and Correction of Human Error", *Journal of Applied Behavioral Sciences*, 32: 5–32.

Edwards, E. (1972). *Man and Machine: Systems for Safety. Proceedings of British Airline Pilots Association Technical Symposium.* London, England, British Airline Pilots Association: 21–36.

Edwards, E. (1975). "Stress and the Airline Pilot". Paper presented to British Airline Pilots Association Technical Symposium: Aviation Medicine and the Airline Pilot. London, England.

Flight Safety Foundation (1999). "Erroneous Airspeed Indications Cited in Boeing 757 Control Loss", *Accident Prevention*, 56 (10): 1–8.

Foushee, H.C., J.K. Lauber, M.M. Baetge and D.B. Acomb (1986). "Crew Performance as a Function of Exposure to High-density, Short-haul Duty Cycles", NASA TM 88322. Moffett Field, CA: NASA-Ames Research Center.

Foushee, H.C. and K.L. Manos (1981). "Information Transfer Within the Cockpit: Problems in Intracockpit Communications", in Billings, C.E. and E.S. Cheaney (eds) *Information Transfer Problems in the Aviation System*. NASA TP 1875. Moffett Field, CA: NASA-Ames Research Center.

Ginnett, R.C. (1987). "First Encounters of the Close Kind: The Formation Process of Airline Flight Crews", Doctoral dissertation, Yale University, CT.

Gröning, M. and P. Ladkin (1999). Bericht der Direccion General de Aeronautica Civil über die Untersuchung des Unfalles mit dem Flugzeug Boeing B-757 am 06. February 1996 bei Puerto Plato (in der Übersetzung vom Luftfahrtbundesamt); Universität Bielefeld; Technische Fakultät, AG Rechnernetze und Verteilte Systeme. http://www.rvs.uni-bielefeld.de/publications/Incidents/DOCS/ComAndRep/PuertoPlata/bericht.html. 15.12.2008.

Guillén, L., K. Korotov and M. Mayo (2012). "Is Leadership a Part of Me?", ESMT Working Paper, 11–04 (R1).

Hackman, J.R. (1986). "Group-Level Issues in the Design and Training of Cockpit Crews", in Orlady, H.W. and H.C. Foushee (eds) *Proceedings of the nasa/mac Workshop of Cockpit Resource Management Training*. NASA CP 2455 (23–39). Moffett Field, CA: NASA-Ames Research Center.

Hackman, J.R. and R.L. Helmreich (1987). "Assessing the Behavior and Performance of Teams in Organizations: The Case of Air Transport Crews", in Peterson, D.R. and D.B. Fisherman (eds) *Assessment for Decision* (283–313). New Brunswick, NJ: Rutgers University Press.

Hackman, J.R. and C.G. Morris (1975). "Group Tasks, Group Interaction Process, and Group Performance Effectiveness: A Review and Proposed Integration", in Berkowitz, L. (ed.) *Advances in Experimental Social Psychology*, Vol. 8. New York, NY: Academic Press.

Hamilton, S. (2004). *The Enron Collapse*. IMD Case Study.

Haynes, A.C. (1991). The Crash of United Flight 232. Dryden Flight Research Facility, Edwards, CA: NASA Ames Research Center. http://yarchive.net/air/airliners/dc10_sioux_city.html. October 05, 2012.

Haynes, A.C. (1991a). "United 232: Coping with the Loss of All Flight Controls", *Air Line Pilot*, October 1991: 10–14, 54–55 and November 1991: 26–28.

Heinzer, T. H. (1993). "Enhancing the Impact of Human Factors Training", in *International Civil Aviation Organization. Proceedings of the Second ICAO Global Flight Safety and Human Factors Symposium*, Washington DC, April 12–15, 1993 (190–196). Montreal, Canada: ICAO.

Helmreich, R.L. (1994). "Anatomy of a System Accident: The Crash of Avianca Flight 052", *The International Journal of Aviation Psychology*, 4 (3): 265–284.

Helmreich, R.L. and H.C. Foushee (1993). "Why Crew Resource Management? Empirical and Theoretical Bases of Human Factors Training in Aviation", in Wiener, E.L., B.G. Kanki and R.L. Helmreich (eds) *Cockpit Resource Management* (3–45). San Diego, CA: Academic Press.

Helmreich, R.L., A.C. Merritt and J.A. Wilhelm (1999). "The Evolution of Crew Resource Management Training in Commercial Aviation", *International Journal of Aviation Psychology*, 9 (1): 19–32.

House of Representatives (1989). Honoring the Crew of United Airlines Flight 811, May 10, 1989. Library of Congress: H 1798.

Janis, I.L. (1972). *Victims Of Groupthink: a Psychological Study of Foreign Policy Decisions and Fiascoes*. Boston, MA: Houghton Mifflin.

Janis, I.L. (1982). *Groupthink: a Psychological Study of Policy Decisions and Fiascoes*. Boston, MA: Houghton Mifflin.

Kanki, B.G. and M.T. Palmer (1993). "Communication and Crew Resource Management", in Wiener, E.L., B.G. Kanki and R.L. Helmreich (eds) *Cockpit Resource Management* (99–136). San Diego, CA: Academic Press.

Kern, T. (1995). *Darker Shades of Blue: A Case Study of Failed Leadership*. United States Air Force Academy, Department of History, Colorado Springs: United States Air Force Academy.

Kern, T. (2001). *Culture, Environment, and CRM – Controlling Pilot Error*. New York, NY: McGraw-Hill.

Kern, T. (2009). *Blue Threat: Why to Err is Inhuman*. Monument, CO: Pygmy Books.

Küpper, M. (2012). *Flughafen-Kosten "unkalkulierbar"*. Frankfurter Allgemeine Zeitung, May 21, 2012.

Ladkin, P. (1999a). Birgenair B757 Accident Intra-Cockpit Communication. Universität Bielefeld; Technische Fakultät, AG Rechnernetze und Verteilte Systeme. http://www.rvs.uni-bielefeld.de/publications/

Incidents/DOCS/ComAndRep/PuertoPlata/CVR-birgenair.html. December 16, 2008.

Ladkin, P. (1999b). Birgenair B757 Accident, March 18, 1996 Press Release (Translation: Released by US NTSB); Universität Bielefeld; Technische Fakultät, AG Rechnernetze und Verteilte Systeme. http://www.rvs. uni-bielefeld.de/publications/Incidents/DOCS/ComAndRep/ PuertoPlata/18-3-birgenair.html. December 16, 2008.

Lutz, T. (2004). "Throttles Only Control (TOC)", *InterPilot* 2004: 32–34.

McKinney, E.A., J.R. Barker, K.J. Davis and D. Smith (2005). "How Swift Starting Action Teams Get Off the Ground", *Management Communication Quarterly*, 19 (2): 198–237.

Milliken, F.J., E.W. Morrison and P.F. Hewlin (2003). "An Exploratory Study of Employee Silence: Issues that Employees Don't Communicate Upward and Why", *Journal of Management Studies*, 40 (6): 1453–1476.

Morgan, L. (1983). "Just Relax", *Flying*, 110 (12): 22–23.

National Transportation Safety Board (1978). Aircraft Accident Report. Japan Air Lines Co., McDonnell-Douglas DC-8–62F, JA 8054, Anchorage, Alaska, January 13, 1977. NTSB-AAR-78–7. Washington, DC: NTSB.

National Transportation Safety Board (1979). Aircraft Accident Report. United Airlines, Inc., McDonnell-Douglas, DC-8–61, N8082U. Portland, Oregon, December 28, 1978. NTSB–AAR-79–7. Washington, DC: NTSB.

National Transportation Safety Board (1980). Aircraft Accident Report. Air New England, Inc., DeHavilland DHC-6, N383EX. Hyannis, Massachusetts, June 17, 1979. NTSB-AAR-80–1. Washington, DC: NTSB.

National Transportation Safety Board (1986). Aircraft Accident Report. Air Canada Flight 797, McDonnell Douglas DC-9–32, C-FTLU, Greater Cincinnati International Airport Covington, Kentucky, June 2, 1983. NTSB/AAR-86/02. Washington, DC: NTSB.

National Transportation Safety Board (1990). Aircraft Accident Report. United Airlines Flight 232, McDonnell-Douglas DC-10–10, Sioux City, Iowa, July 19, 1989. NTSB/AAR-90/06. Washington, DC: NTSB.

National Transportation Safety Board (1991). Aircraft Accident Report. Avianca, The Airline of Columbia, Boeing 707–321B, HK 2016, Fuel Exhaustion, Cove Neck, New York, January 25, 1990. NTSB/AAR-91/04. Washington, DC: NTSB.

National Transportation Safety Board (1992). Aircraft Accident Report. Explosive Decompression – Loss of Cargo Door in Flight, United

Airlines Flight 811, Boeing 747–122, N4713U, February 24, 1989. NTSB/AAR-92/02. Washington, DC: NTSB.

National Transportation Safety Board (1994). "A Review of Flightcrew-Involved, Major Accidents of U.S. Air Carriers, 1978 through 1990, Safety Study", NTSB SS-94/01. Washington, DC: NTSB.

National Transportation Safety Board (2000). Aircraft Accident Report. In-flight Breakup Over the Atlantic Ocean, Trans World Airlines Flight 800, Boeing 747–131, N93119, Near East Moriches, New York, July 17, 1996. NTSB/AAR-00/03. Washington, DC: NTSB.

Netherlands Aviation Safety Board (1977). Final Report and Comments of the Investigation into the Accident with the Collision of KLM Flight 4805, Boeing 747–206B, PH-BUF and Pan American Flight 1736, Boeing 747–121, N736PA at Tenerife Airport, Spain on 27 March 1977. ICAO Circular 153-AN/56.

Noelle-Neumann, E. (1974). "The Spiral of Silence: A Theory of Public Opinion", *Journal of Communication*, 24 (2): 43–51.

Oppel, R.A. (2002). *Enron Official Says Many Knew about Shaky Company Finances*. New York Times, 15 February 2002.

Orlady, H.W. and H.C. Foushee (1987). "Cockpit Resource Management Training", NASA CP 2455. Moffett Field, CA: NASA-Ames Research Center.

Pariès, J. and R. Amalberti (1995). "Recent Trends in Aviation Safety: From Individuals to Organizational Resources Management Training", Risøe National Laboratory Systems Analysis Dept. (eds) *Technical Report, Risøe Series 1* (216–228). Roskilde, Denmark: Risøe National Laboratory.

Perrow, C. (1999). *Normal Accidents: Living with High-Risk Technologies*. Princeton, NJ: Princeton University Press.

Predmore, S. (1991). "Microcoding of Communications in Accident Investigation: Crew Coordination in United 811 and United 232", in Jensen, R.S. (ed.) *Proceedings of the Sixth International Symposium on Aviation Psychology* (350–355). Columbus, OH: The Ohio State University.

Rall, M., B. Schaedle, J. Zieger, W. Naef and M. Weinlich (2002). "Neue Trainingsformen und Erhöhung der Patientensicherheit", *Unfallchirurg*, 11 (105): 1033–1042.

Reason, J. (1990). *Human Error*. New York, NY: Cambridge University Press.

Reason, J. (1997). *Managing the Risks of Organizational Accidents*. Aldershot, England: Ashgate Publishing.

Redding, W.C. (1985). "Rocking Boats, Blowing Whistles, and Teaching Speech Communication", *Communication Education*, 34: 245–258.

Richthofen, M.F.v. (1917). *Der rote Kampfflieger*. Berlin, Germany: Ullstein.

Roberts, K.H. and C.A. O'Reilly (1974). "Failures in Upward Communication in Organizations: Three Possible Culprits", *Academy of Management Journal*, 17: 205–215.

Roitsch, P.A., G.L. Babcock, and W.W. Edmunds (1979). *Human Factors Report on the Tenerife Accident*. Washington, DC: Airline Pilots Association.

Rosay, J. (2004). "Baghdad A300 Incident Discussion", Airbus Presentation, October 28, 2004.

Rosen, S. and A. Tesser (1970). "On Reluctance to Communicate Undesirable Information: The MUM Effect", *Sociometry*, 33: 253–263.

Ruffell Smith, H.P. (1979). "A Simulator Study of the Interaction of Pilot Workload with Errors, Vigilance, and Decisions", NASA TM 748482. Moffett Field, CA: NASA-Ames Research Center.

Rutherford, M. (no date). Eyewitness Report: United Flight 811. http://www.airdisaster.com/eyewitness/ua811.shtml December 16, 2008.

Schulz, K. (2010). *Being Wrong – Adventures in the Margin of Error*. New York, NY: HarperCollins Publishers.

Sexton, J.B., E.J. Thomas and R.L. Helmreich (2000). "Error, Stress, and Teamwork in Medicine and Aviation: Cross Sectional Surveys", *BMJ*, 320 (7237): 745–749.

Snook, S.A. (2000). *Friendly Fire – The Accidental Shootdown of U.S. Black Hawks over Northern Iraq*. Princeton, NJ: Princeton University Press.

Soane, E., N. Nicholson and P. Audia (1998). "The Collapse of Barings", LBS Case Study 401–020–1 und 401–021–1.

Stonham, P. (1995). "Whatever Happened at Barings?", EAP Case Study 295–033–1 und 295–034–1.

Sturbeck, W. (2012). *Thyssen-Krupp stellt Werke zum Verkauf*. Frankfurter Allgemeine Zeitung, 15 May 2012.

Transcript Cockpit Voice Recorders UAL Flight 811, N4713U (1990). http://www.planecrashinfo.com/cvr890224.htm 16.12.2008.

Tri-City-Herald (1989). *Air Controller Remembers Moments of Hope*. August 19, 1989: C-3.

US Air Force (1995). AFR 110–14 USAF Accident Investigation Board. 24 June 1994, Fairchild AFB, WA, B-52H Aircraft S/N 61-0026 92 BW 325 BS. HQ 12th Air Force. Obtained under the Freedom of

Information Act on April 30, 2009 by the Department of the Air Force, Headquarters Air Combat Command. Langley Air Force Base, VA.

Weener, E.F. (1992). "Action Must Be Taken to Further Reduce the Current Accident Rate as the Transport Fleet Increases in Size and Operation", *Accident Prevention*, 49 (6): 1–8. Alexandria, VA: Flight Safety Foundation.

Weick, K.E. (1990). "The Vulnerable System: An Analysis of the Tenerife Air Disaster", *Journal of Management*, 16 (3): 571–593.

Zao, B. and F. Olivera (2006). "Error Reporting in Organizations", *Academy of Management Review*, 31 (4): 1012–1030.

Printed and bound by CPI Group (UK) Ltd, Croydon, CR0 4YY